MadCap Flare 2023: The De

Scott DeLoach

ClickStart, Inc.
www.clickstart.net

Designer: Patrick Hofmann

Developed in MadCap Flare

Dedication

This book is dedicated to my mother, Sylvia Jo DeLoach.
Thanks, mom, for always believing in me.

Contents

Importing Markdown files 102

Importing external resources.....................105

Importing content from Flare projects 107

Lists..111

Tables 114

Images and multimedia 129

Single source 297

Condition tags 313

Micro content 323

Build and publish 329

Targets 331

Source control with Central 441

SharePoint integration 446

Accessibility 465

Internationalization.................. 473

Flare customization 481

Appendices 493

Additional resources 495

Keyboard shortcuts (by task) 496

Introduction

This book was designed to be both a study guide for the certified MAD program and a comprehensive guide for Flare users. It meets the needs of a wide range of Flare users, including those who are:

- ☐ New to developing content for knowledge bases, help systems, user guides, learning or training solutions, or policies & procedures

- ☐ Migrating from Author-it

- ☐ Migrating from Confluence

- ☐ Migrating from Doc-to-Help

- ☐ Migrating from RoboHelp

- ☐ Migrating from FrameMaker

- ☐ Migrating from Word

- ☐ Developing content for use in Salesforce, ServiceNow, SharePoint, or Zendesk

- ☐ Upgrading from a previous version

- ☐ Looking for a quick "refresher" of key features

- ☐ Preparing to be Certified MAD for Flare™

This book provides the essential information you need to use all of Flare's major features. It also includes information that will appeal to advanced users, such as keyboard shortcuts, Flare file descriptions, and a quick task index.

What is 'Certified MAD for Flare?'

The certified MadCap Advanced Developer (MAD) program recognizes and validates your ability to use MadCap products. Being certified MAD for Flare is the best way to demonstrate your abilities and stay up to date with MadCap Flare. It also sounds cool!

To be certified MAD for Flare, you will need to:

- pass the certification test
- submit a sample Flare project

About the exam

The certification test is a 75-minute, 50-question test that you take on the Web. The passing score is 70%. The test is not easy—you will need to study and review this book's sample questions to pass. We designed the certification program to be a true assessment of Flare users' advanced abilities, and we want those who pass to be proud to be Certified MAD for Flare!

About the sample project

The sample project is a Flare project that demonstrates your ability to use Flare. The project does not have a time limit, and you can choose the subject.

Preparing for certification

The best way to prepare for the certification test is to take a Flare class. The questions on the test are drawn directly from the course guides and class content. You will also learn best practices for creating Flare projects, which will help you successfully pass the sample project requirements.

This guide is also a great way to prepare for certification, and it is designed to complement the training classes. The questions at the end of each chapter are similar to the test questions, and they can be used to review Flare features before taking the test. The step-by-step instructions will help you successfully create the sample project.

Icons used in this guide

The following icons are used throughout this guide to help you find important and time-saving information.

Icon	Meaning	Description
⚠	Caution	Important advice that could cause data loss or unnecessary aggravation if not followed.
NEW!	New Feature	A new or substantially enhanced feature in the latest version of Flare.
◇	Note	Additional information about a topic.
TIP▶	Tip	A recommended best practice, shortcut, or workaround.

Updates

For the most up-to-date information about this book, see **www.clickstart.net**.

For the most up-to-date information about Flare, see MadCap Software's website at **www.madcapsoftware.com** and the Flare forums at **forums.madcapsoftware.com**.

What's new in Flare 2023

Flare 2023 includes several user-requested enhancements and bug fixes. To review the full list, see
kb.madcapsoftware.com/Default.htm#CSHID=GEN1074F

The major new and enhanced features of Flare 2023 are listed in the table below.

Feature	See Page
Branding stylesheets	33, 260
Using CSS variables for colors in table stylesheets and skins	238
Sending review packages for review in Central	416
Support for MadCap Central servers in Europe	441

Six reasons to use Flare

Flare is an advanced XML-based content authoring application with powerful single sourcing features. In addition to its innovative user interface and excellent online help, Flare has many strengths that make it a great choice for developing knowledge bases, online help, policies and procedures, user guides, training materials, and technical manuals. This section lists six reasons why I use Flare and recommend it to clients.

XML-based architecture and clean code

Flare's XML-based architecture allows MadCap to support multiple XML schemas such as XHTML, Markdown, and DITA and potentially add support for additional schemas in the future. All of Flare's project files are XML-based, so they're extremely small and easy to read in Notepad or an XML editor.

Flare's XML-based authoring also means that it produces clean code.

Content linking

Flare allows you to import Microsoft Word, Microsoft Excel, Adobe FrameMaker, Confluence, HTML, XHTML, Markdown, and DITA documents and content from other Flare projects. When you import content, you can link the imported topics to the source document or Flare project.

Content linking allows you to maintain your content in different applications and reuse it in Flare. For example, your coworkers can develop content in Word and you can import their content into Flare. If you have multiple Flare projects, you can link common formatting elements such as stylesheets and page layouts and reuse them.

Page layouts

Page layouts can be used to set the page size and margins and to set up headers and footers for print targets. Flare's page layouts are very advanced: you can set up different headers and footers for title, first, empty, odd, and even pages, and you can create and use multiple page layouts. For example, you can use a landscape page layout for wide topics and a two-column page layout for your index.

Snippets

Snippets can be used to reuse any content, including text, images, and tables in multiple topics. You can use snippets to reuse a note, a procedure, or even a screenshot and its description.

Source control support

Because Flare has an open XML architecture, Flare projects are compatible with most source control applications. Flare provides integrated support for Git, Perforce Helix Core, Subversion (SVN), and Team Foundation Server (TFS), which means you can check files in or out and perform other source control tasks from within Flare. You can even set up Flare to automatically send an email or instant message to another team member if you need to modify a file they have checked out.

Table styles

In other help authoring tools and HTML editors, you must use inline formatting to format tables. If you need to change the table formatting, it's usually very tedious and time consuming.

In Flare, you can use table styles to specify table borders, background colors, captions, and other properties. You can even format header and footer rows and set up alternating background colors for rows and columns. Flare allows you to create multiple table styles, so you can create online- and print-specific table styles.

Projects

This section covers:

- [ ] Creating a new project
- [ ] Converting from Author-it
- [ ] Converting from Confluence
- [ ] Converting from Doc-to-Help
- [ ] Converting from RoboHelp
- [ ] Importing HTML Help files
- [ ] Converting from FrameMaker

New projects

Project files have a .flprj extension (for "Flare project"). Flare's project file is a small XML file—feel free to open it in Notepad and take a look.

The project file is stored in your project's top-level folder. You name this folder when you create a new project. For example, if your project is named "MyFirstProject," your top-level folder is named "MyFirstProject." By default, Flare creates your top-level folder in the My Documents\My Projects folder.

In addition to the project file, your top-level folder contains four subfolders: Analyzer, Content, Project, and Output.

The **Analyzer** folder contains data for the project analysis reports, such as broken links and topics that are not in a TOC or index.

The **Content** folder contains your topics, images, sounds, stylesheets, and movies.

The **Project** folder contains your conditional tag sets, context-sensitive help map files, glossaries, skins, TOCs, and variable sets. Wondering where the index file is? There's not one—Flare stores your index keywords in your topics.

The **Output** folder contains your generated targets, like HTML5 or PDF.

'Which languages does Flare support?'

Flare provides full support for the following languages:

□	Arabic	□	Finnish	□	Norwegian
□	Chinese	□	French	□	Persian
□	Danish	□	German	□	Portuguese
□	Dutch	□	Hebrew	□	Spanish
□	English	□	Italian	□	Swedish
□	Thai	□	Urdu	□	Yiddish

Flare supports Unicode, so you can write your topics in any language. Flare also includes translated skins for the languages listed above and for numerous region-specific languages. For more information, see "Selecting a language for a target" on page 476.

The Flare interface can be viewed in English, Chinese, French, German, or Japanese. You can change the language in the Select UI Language dialog box when you open Flare. If the dialog box does not appear, you can turn it on by selecting **File > Options** and selecting the **Show Select UI Language Dialog on Startup** option on the General tab.

Creating a new project NEW!

You can create a new project using a template. In fact, you use templates to create everything in Flare, including topics, stylesheets, glossaries, skins, and variable sets. Project templates include sample topics, page layouts, template pages, stylesheets, and other files to give you a head start designing your project. If you have your own design files, you can use the "Empty" template.

In Flare 2023 and later, you can also set up CSS variables in a branding stylesheet to specify your company's colors and logo.

Shortcut	Tool Strip	Ribbon
Alt+F, N	(Standard toolbar)	File > New Project > New Project

To create a new project:

1   Select **File > New Project > New Project**.
    The Start New Project wizard appears.

2 Type a **Project Name**.

You don't have to type the .flprj extension. Flare will add it for you if you leave it out.

3 Type or select a **Project Folder**.

4 Select a **Language** and click **Next**.

The language you select determines which dictionary is used for spell checking.

5 Select a **Source** and click **Next**.

You can select a factory template or your own project template.

6 If needed, customize the design elements for your project's branding stylesheet and click **Next**.

7 Select an **Available Target** and click **Finish**.

Your new project opens in Flare.

Converting from Author-it

Importing an Author-it project

You can use the MadCap Author-it® to Flare Converter to create a new Flare project based on Author-it files.

To install the Author-it converter:

1 Download the Author-it Converter from the following link: www.madcapsoftware.com/downloads/plugins-api-redistributables.aspx

2 Right-click the AuthorItConverterSetup.exe and select **Run as administrator**.

To import an Author-it project:

1 Publish your Author-it project to XML.

2 Open the MadCap Author-it to Flare Converter.

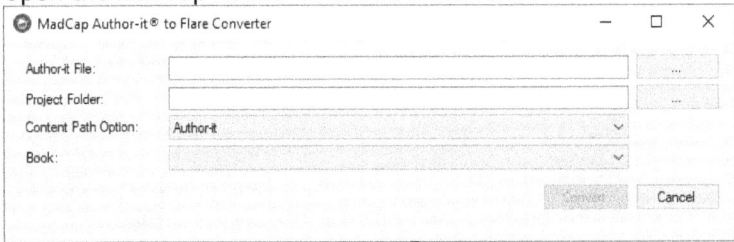

3 Locate and select your published XML file from Author-it.

4 Select a **Project Folder**.

5 Select a **Content Path Option**:

 □ **Author-it** – maintain the existing folder structure

 □ **Flat** – add all topics to the Content folder

 □ **Condensed** – maintain the existing folder structure and remove any empty folders

6 If your Author-it project contains multiple books, select a **Book**.

7 Click **Convert**.

'What happens to my... ?'

The following table explains how Author-it's features convert to Flare.

Author-it Feature	Converts?	Comments
Context IDs	✓	Converted to alias and header files.
Embedded topics	✓	Converted to snippets and stored in the Resources folder.
Glossaries	✓	Converted to glossary and glossary proxy.
Images	✓	Stored in the Resources folder.
Index keywords	✓	
Stylesheets	✓	Stored in the Resources folder.
Tables of contents	✓	Converted to topics.
Variables	✓	Variables used in variables become alternate variable definitions.

Converting from Confluence

You can create a new project based on Confluence pages, or you can import Confluence pages into an existing Flare project. If you link the imported topics to their source pages, you can edit the Confluence pages and re-import them into your project.

Creating a new project based on Confluence pages

When you import Confluence pages, Flare saves your settings in a Confluence Import File (these files have a .flimpconf extension). You can reuse these settings when you re-import the Confluence pages or import similar pages.

If you want to import Confluence pages into an existing project, see "Importing Confluence pages" on page 39.

Shortcut	Tool Strip	Ribbon
none	Project > Import Project > DITA Document Set	File > New Project > DITA Document Set

To create a new project based on Confluence pages:

1   Select **File** > **New Project** > **Confluence Pages**.
    The Import Confluence wizard appears.

2 Type a **Project Name**.

3 Type or select a **Project Folder**.

4 Select an **Output Type**.
You can add additional output types to your project later.

5 Type the path to the **Confluence Server**.

6 Type your **Username**.

7 Type your **Password** (for Confluence Server or Data Center) or **API Token** (for Confluence Cloud).

8 Click **Submit**.
The Workspace Selection page appears.

9 Select a **Space**.

10 Select the pages you want to import.

11 Select the **Advanced Options** tab.

12 Select **Import linked pages** if you also want to import any pages that are linked to the selected pages.

13 Select **Remove inline formatting** in you do not want to keep any inline formatting that has been applied to your content.

14 Select **Remove style classes** if you want to remove any style classes that Confluence has applied to your content.

15 Select **Import resources** if you want to import any images or multimedia files that are used in your pages.

16 If you plan to continue editing the original Confluence pages, select **Link generated files to source files**.

◇ *This option allows you to link the imported topics to the source Confluence pages. When you re-import the pages, Flare will replace the original topics with the new topics.*

Linked topics have a chain (⊘) icon after their file name when opened in the XML Editor.

17 Click **Finish**.

Importing Confluence pages into a Flare project

You can import Confluence pages into an existing Flare project. Each page becomes a Flare topic.

Shortcut	Tool Strip	Ribbon
Alt+P, I, C	Project > Import File > Add Confluence Import File	File > New

To import Confluence pages:

1 Select **File** > **New**.
—OR—
Right-click the **Imports** folder and select **Add Confluence Import File**.
The Add File dialog box appears.

2 For **File Type**, select **Confluence Import File**.

3 Select a **Source** template.

4 Type a **File Name**.

5 Click **Add**.
   The Confluence Import Editor appears.

6 For **Import to folder**, type or select a folder for the imported topics.

7 Type the path to the **Confluence Server**.

8 Type your **Username**.

9 Type your **Password** (for Confluence Server or Data Center) or **API Token** (for Confluence Cloud)

10 Select the **Workspace Selection** tab.

11 Select a **Space**.

12 Select the pages you want to import.

13 Select the **Advanced Options** tab.

14 Select **Import linked pages** if you also want to import any pages that are linked to the selected pages.

15 Select **Remove inline formatting** in you do not want to keep any inline formatting that has been applied to your content.

16 Select **Remove style classes** if you want to remove any style classes that Confluence has applied to your content.

17 Select **Import resources** if you want to import any images or multimedia files that are used in your pages.

18 If you plan to continue editing the original Confluence pages, select **Link generated files to source files**.

◇ *This option allows you to link the imported topics to the source Confluence pages. When you re-import the pages, Flare will replace the original topics with the new topics.*

Linked topics have a chain (⊗) icon after their file name when opened in the XML Editor.

19 Click **Import** in the toolbar.
The Accept Imported Documents dialog box appears.

20 Click **Accept**.
The imported topics appear in the Content Explorer in the folder you selected on the General tab.

'What happens to my... ?'

The following table explains how Confluence's features convert to Flare.

Confluence Feature	Converts?	Comments
Paragraph styles		
Article title	✓	h1
Heading 1 through Heading 6	✓	

Confluence Feature	Converts?	Comments
Preformatted	✓	
Quote	✓	Converted to blockquotes.
Character styles		
Bold	✓	
Font colors	✓	Converted to span tag with the color set using inline formatting.
Italic	✓	
Monospace	✓	Converted to code tags.
Strikethrough	✓	
Subscript	✓	
Superscript	✓	
Underline	✓	
Lists		
Bulleted list	✓	
Numbered list	✓	
Task list	✓	Converted to bulleted lists.
Formatting		
Code block	✓	Converted to div tags.
Column	✓	Converted to div tags.
Expand	✓	Converted to div tags.
Info	✓	Converted to div tags.
No format	✓	Converted to div tags.
Note	✓	Converted to div tags.
Panel	✓	Converted to div tags.
Section	✓	Converted to div tags.
Tip	✓	Converted to div tags.
Warning	✓	Converted to div tags.

Converting from Doc-to-Help

Importing a Doc-to-Help project

You can import projects created with Doc-to-Help (.d2h files) into Flare.

⬦ Since Flare is a 64-bit application, you'll need to download Microsoft Access Database Engine 2010 x64 edition. If you don't already have the database engine, Flare will prompt you to download it.

Shortcut	Tool Strip	Ribbon
None	File > Import Project > Doc-to-Help Project	File > New Project > Doc-to-Help Project

To import a Doc-to-Help project:

1 Select **File** > **New Project** > **Doc-to-Help Project**.
   The Open File dialog box appears.

2 Locate and select a Doc-to-Help .d2h project file.

3 Click **Open**.
   The Import Project Wizard appears.

4 Click **Next**.

5 Type a **Project Name**.

6 Type or select a **Project Folder** and click **Next**.

7 Click ⬚.

8 Select the folder where you store your CSS files (stylesheets) and click **Select Folder**.

9 Click **Next**.

10 Select a **Language** for spell checking and click **Finish**.
   The Doc-to-Help project is imported into Flare. Your new Flare project file will have a .flprj extension.

'What happens to my... ?'

The following table explains how Doc-to-Help's features convert to Flare.

Doc-to-Help Feature	Converts?	Comments
Attributes	✓	Converted to condition tags.
Bookmarks	✓	
Collapsible sections	✓	Converted to togglers.
Comments	✓	Converted to annotations.
CSS	✓	
Drop-down links	✓	
Expanding links	✓	
Glossaries	✓	Converted to glossary and glossary proxy.
Groups	✓	Converted to concepts and concept links.
Images	✓	
Index keywords	✓	
Pop up links	✓	Converted to text popups.
Link tags	✓	Converted to bookmarks.
Related topics links	✓	
Targets	✓	Converted, but the settings do not import.
Themes	✓	Converted to skins.
TOCs	✓	
Variables - plain text	✓	Converted to variables.
Variables - rich content	✓	Converted to snippets.
Videos	✓	

Doc-to-Help Feature	Converts?	Comments
Widgets - carousel	✓	Converted to slideshows.
Widgets - CodeHighlighter	✓	Converted to div tags.
Widgets - gallery	✓	Converted to slideshows.
Widgets - lightbox	✓	Converted to slideshows.
Widgets - note	✓	Converted to div tags.
Widgets - tabs	✓	Converted to div tags.
Widgets - topic content	✓	Converted to mini-TOC proxies.

Converting from RoboHelp

Top ten RoboHelp conversion 'gotchas'

RoboHelp and Flare are similar, but there are some features that are *just* different enough to be confusing. There are also a few Flare features that can be hard to find, especially if you are accustomed to using RoboHelp.

Here's my list of the top ten features that I had trouble understanding, finding, and remembering how to use when I started using Flare.

10 Panes vs panels

RoboHelp's UI contains "panels," such as the Contents, Properties, and Table of Contents panels. In Flare, these UI elements are called panes, and they often have slightly different names or are organized differently. RoboHelp organizes panels into Author and Output views, which (in general) match Flare's Content Explorer and Project Organizer tabs.

9 Using condition tag boxes

Flare's Content Explorer provides condition tag boxes to identify topics that use condition tags. Unfortunately, an empty condition tag box looks like a checkbox. I tried to select these boxes for a few days and thought they must be broken. According to the Flare help community, many other users assume they are checkboxes too.

8 Adding image thumbnails

In RoboHelp, you can set an image to display as a smaller "thumbnail" icon in online targets. Users can then click the thumbnail icon to view the full-sized image.

In Flare, you can use the mc-thumbnail style property to automatically add thumbnail icons for images that are larger than a selected width or height.

7 Auto-creating a TOC

In RoboHelp, you can auto-create a TOC based on your Contents panel. Folders in the Contents panel become books, and topics become pages. If you change the organization in the Contents panel, you can recreate the TOC or update it yourself.

In Flare, you can use drag-drop to add folders and/or topics to your TOC. You can also set up a TOC book to automatically create links to headings inside a topic. When you add or remove headings in the topic, the TOC is automatically updated. You can also use a variable to automatically use a topic's heading as the TOC book or page's label. See "Creating TOC books and pages" on page 187.

6 Using template pages

RoboHelp's master pages are similar to Flare's template pages, but Flare's template pages are only used for online targets (print targets use page layouts). RoboHelp's master pages are used for online and print targets.

5 Using page layouts

Flare's page layouts can be used to set the page size, page margins, headers, and footers for print documents. In RoboHelp, these settings are specified in a master page and in a PDF output preset.

4 Modifying skins

RoboHelp and Flare both use skins to design online outputs. However, the design options are very different. You will need to recreate your RoboHelp skins when you import a RoboHelp project into Flare.

3 RoboHelp's HTML5 output presets vs Flare's HTML5 targets

RoboHelp's output presets and Flare's targets are both used to create outputs. However, they have different default and optional settings. After you import, you should review the settings in your output presets to adjust them as needed and to take advantage of Flare's output options.

2 Viewing browse sequences

In RoboHelp, browse sequences appear as either a graphical bar at the top of your topics (HTML Help) or as small arrows in the navigation pane (HTML5 and Frameless). RoboHelp's HTML Help browse sequences require the HHActiveX.dll file to be installed on the user's computer.

In Flare, browse sequences appear as either a TOC item (HTML Help) or as an accordion item (other online target types such as HTML5). Because they do not appear on a custom tab, Flare's HTML Help browse sequences do not require a .dll file.

You can add browse sequence next and previous buttons to your toolbar using a skin. For more information, see "Using a browse sequence" on page 227.

1 Context-sensitive help paths

Flare organizes your generated HTML5 and WebHelp topics in a "Content" folder. Since RoboHelp does not use a Content folder, your context-sensitive help links might not work after you convert to Flare. If you cannot change the code to include the Content folder, you can remove it from your HTML5 and WebHelp files by selecting "Do not use 'Content' folder in output" on the Advanced tab in your target.

Importing a RoboHelp project

You can import projects created with RoboHelp X5 or later (.xpj files) or RoboHelp X4 or earlier (.mpj files). If you want to import an HTML Help project created with another help authoring tool, see "Importing an HTML Help file" on page 52.

Shortcut	Tool Strip	Ribbon
none	File > Import Project > RoboHelp Project	File > New Project > RoboHelp Project

To import a RoboHelp project:

1 Select **File** > **New Project** > **RoboHelp Project**.
The Open File dialog box appears.

2 Locate and select a project file.

- ☐ RoboHelp X5 or later: .xpj file

- ☐ RoboHelp X4 or earlier: .mpj file

3 Click **Open**.
The Import Project Wizard appears.

4 Click **Next**.

5 Type a **Project Name**.

6 Type or select a **Project Folder** and click **Next**.

7 Select whether you want to **Convert all topics at once**.
This option converts your topic files from HTML to XHTML. If
you don't select this option, your files will remain as HTML files
and your index terms will not be imported.

8 Select whether you want to **Convert inline formatting to CSS
styles**.
⚠ *Be careful with this option. If you use a lot of inline
formatting, you could end up with hundreds of styles.*

9 Click **Next**.

10 Select a **Language** for spell checking and click **Finish**.
The RoboHelp project is imported into Flare. Your new Flare
project file will have a .flprj extension.

'What happens to my... ?'

The table on the following page explains how RoboHelp's features
convert to Flare.

RoboHelp Feature	Converts?	Comments
Browse sequences	✓	
Conditional tags	✓	Stored in a condition tag set named "Primary."
Custom colors		You must recreate the colors in Flare.

RoboHelp Feature	Converts?	Comments
Dictionaries		You can add your terms to a global or project-specific dictionary. See "Adding spell check dictionaries" on page 388.
HTML topics	✓	Converted to XHTML either when imported (recommended) or when opened.
Folders	✓	
Forms	✓	Forms are converted, but you can only edit them in the Text Editor.
Frames	✓	Frames are converted, but you can only edit them in the Text Editor.
Glossary	✓	
Inline formatting	✓	Maintained, or you can convert them to styles.
Images	✓	
Publishing locations	✓	Converted to "Publishing destinations."
Single-Source layouts	✓	Converted to targets.
Skins		See "Where are my skins?" on page 51.
Snippets	✓	
Sounds	✓	
Table of contents	✓	
Template headers and footers	✓	Converted to snippets.
Variables	✓	Stored in a variable set named "Primary."
Videos	✓	
Windows	✓	Converted to skins.

'Where are my skins?'

RoboHelp skins are imported into Flare, but they are not set up. You can use Flare's Skin Editor to set up your skins.

Importing HTML Help files

You can import an HTML Help .hhp project file or a compiled HTML Help .chm file. If you have the source files and the .hhp file, they will import better. If not, Flare can recreate the source files by decompiling the .chm file.

Creating a new project based on an HTML Help file

You can create a new project based on an HTML Help CHM or HHP file. If you have a CHM file and the source HHP file, you should import the HHP file.

If you want to import a RoboHelp project, see "Importing a RoboHelp project" on page 48.

Shortcut	Tool Strip	Ribbon
none	File > Import Project > HTML File Set	File > New Project > HTML File Set

To create a new project based on an HTML Help file:

1 To import an HTML Help project file, select **File** > **New Project** > **HTML Help Project (HHP)**.
    To import an HTML Help file, select **File** > **New Project** > **HTML Help File (CHM)**.
    The Open File dialog box appears.

2 Locate and select an HHP or CHM file.

3 Click **Open**.
    The Import Project Wizard appears.

4 Click **Next**.

5 Type a **Project Name**.

6 Type or select a **Project Folder** and click **Next**.

7 Select whether you want to **Convert all topics at once**. This option converts your topic files from HTML to XHTML. If you don't select this option, your files will remain as HTML files and your index terms will not be imported.

8 Select whether you want to **Convert inline formatting to CSS styles**.

⚠ *Be careful with this option. If you use a lot of inline formatting, you could end up with hundreds of styles.*

9 Click Next.

10 Select a **Language** for spell checking and click **Finish**. The HTML Help project is imported into Flare.

Converting from FrameMaker

Top ten FrameMaker conversion 'gotchas'

Flare and FrameMaker have very different interfaces. However, both applications can be used to create print documents. When FrameMaker is combined with RoboHelp, you can also use it to create HTML5 targets.

If you focus on tasks and features, FrameMaker and Flare are similar applications. It takes time to learn the Flare interface when you transition, just as it takes time to learn FrameMaker. I've created a list of the top ten differences between FrameMaker and Flare to hopefully make the transition easier for you.

10 Keyboard shortcuts

Like FrameMaker, Flare provides extensive keyboard shortcuts (see page 496 for a list). However, Flare uses "Alt" key shortcuts rather than "Esc" key shortcuts. You can change the default Flare keyboard shortcuts, if needed.

9 Importing content

You can import content into a FrameMaker document, or you can copy and paste content into FrameMaker. In Flare, you can import FrameMaker documents into your project to create new topics. If you want to import content into a topic, you can copy and paste.

8 Inserting graphics

In FrameMaker, images are placed inside anchored or unanchored frames. Placing an image in an anchored frame allows the image to "move" with the surrounding text. Placing an image inside an unanchored frame fixes its position on the page.

In Flare, images automatically "move" with the surrounding text. If you place an image inside a div tag (similar to an anchored frame), you can fix its position.

7 Cross references and hyperlinks

If you import a FrameMaker document that contains cross references, the cross references convert to Flare cross references. In Flare, you can use hyperlinks or cross references to link to other topics. You can keep using cross references to link to topics, but you will need to use hyperlinks to link to websites or other files like PDF documents.

6 Master pages

In FrameMaker, a master page is used to specify the page layout. In Flare, a template page is used to add content to topics in online targets. For example, a Flare template page can be used to add a copyright statement to the bottom of every topic in an HTML5 target.

5 Templates and page layouts

A FrameMaker template contains master pages that specify the formatting, header, and footer for different types of pages (for example, title, odd, and even pages). In Flare, a page layout contains pages that are used to format different types of pages.

4 Lists

In FrameMaker, bulleted and numbered lists are paragraph styles. When you import a FrameMaker document, these list styles remain paragraph styles, just like in FrameMaker. In Flare, lists can also be created using the ul ("unordered" or bulleted) and ol ("ordered" or numbered) list tags rather than the paragraph tag. You can create lists using either method, but it might be confusing to use both approaches in the same project.

3 'Missing' .book files?

Flare does not use .book files. When you import a FrameMaker .book file, Flare will import the included .fm files and create a table of contents (TOC) based on your FrameMaker "TOC" file.

2 Styles

Flare uses styles to format your content. Styles are stored in stylesheets and are basically a combination of the paragraph and character designers in FrameMaker. When you import a FrameMaker document, Flare can create a stylesheet that includes your FrameMaker character and paragraph styles. Table styles can also be converted into table-specific stylesheets.

1 Topic-based authoring

In Flare, your content is separated into short (usually 1-4 printed pages) topics rather than long chapter or section documents. It seems weird and unnecessary at first—why do you need so many small topics? The reason small topics are useful is the same reason multiple chapter documents are useful: it's easier to work with focused "chunks" of content.

In Flare, topics are organized into folders in the Content Explorer and books in the table of contents (TOC). The folders and books are similar to your chapters, and the TOC itself is similar to a FrameMaker .book file. You can easily move topics around in a TOC book, just as you can move .fm chapter documents in a FrameMaker book. You can also easily add new topics to a TOC book. Topic-based authoring also makes it very easy to reuse content in multiple topics. For example, you can reuse a "Copyright" or "Conventions in this guide" topic in multiple Flare projects.

Importing a FrameMaker document

When you import a FrameMaker document, you can divide the document into smaller topics based on styles that are used in your FrameMaker document. For example, you can create a new topic for each "Heading 1" in the document.

Flare can create a stylesheet (.css file) based on the formatting in your FrameMaker document. Or, you can apply an existing stylesheet to reformat your imported topics to match your other topics.

Creating a new project based on a FrameMaker document

You can create a new project based on a FrameMaker document. If you want to import a FrameMaker document into an existing project, see "Importing a FrameMaker document" on page 59.

◇ *When you import a FrameMaker document, Flare saves your settings in a FrameMaker Import File (these files have a .flimpfm extension). You can reuse these settings when you re-import the FrameMaker document or import similar documents.*

Shortcut	Tool Strip	Ribbon
none	File > Import Project > FrameMaker Documents	File > New Project > FrameMaker Documents

To create a new project based on a FrameMaker document:

1   Select **File** > **New Project** > **FrameMaker Documents**.
    The Import FrameMaker wizard appears.

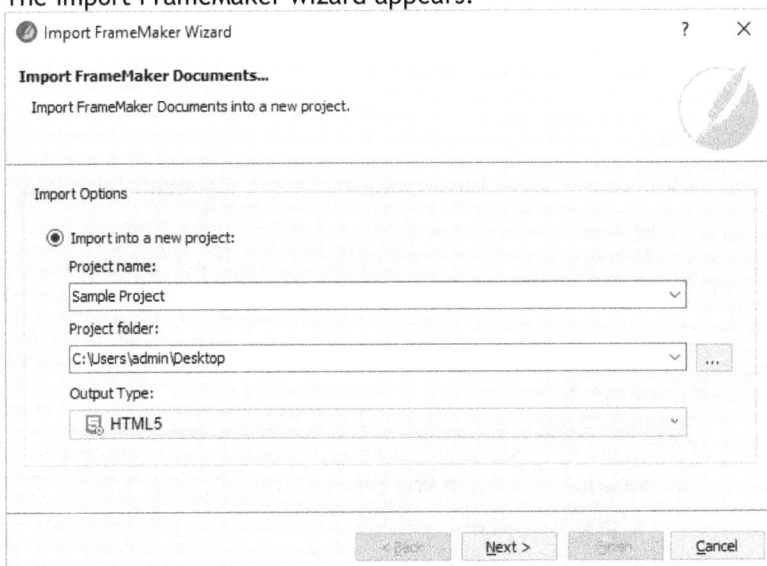

2   Type a **Project Name**.

3   Type or select a **Project Folder**.

4 Select an **Output Type**.
   You can add additional output types to your project later.

5 Click **Next**.

6 Click .

7 Select a .fm or .book FrameMaker file and click **Open**.
   If needed, you can select more than one FrameMaker
   document.

8 If you plan to continue editing the file in FrameMaker, select
   Link generated files to source files.

   ◇ *This option allows you to link the imported topics to the
   FrameMaker document. When you re-import the document,
   Flare will replace the original topics with the new topics.*

   *Linked topics have a chain (⚭) icon after their file name when
   opened in the XML Editor.*

9 Select a style or styles to use to create new topics.
   For example, you can create a new topic for each Heading 1 in
   the FrameMaker document.

10 Click **Next**.

11 If your images have callouts, select **Generate Images with
   Callouts**.
   This option will create a MadCap Capture "props" file for the
   callouts, and you can edit the image and callouts in Capture.

12 If you have sized your images in FrameMaker, select **Preserve
   Image Size**.

13 If you want to automatically reimport the FrameMaker
   document(s) when you create a target, select **Auto-reimport
   before 'Generate Output.'**

14 If your document contains equations, select **Convert equations
   to MathML**.

15 Click **Next**.

16 Select a stylesheet for the new topic(s).

If you select a stylesheet, Flare will apply the stylesheet to your topics and automatically "map" styles with matching names. If you do not select a stylesheet, Flare will create a stylesheet based on the formatting in your FrameMaker document.

17 Click **Next**.

18 Map (or "match") your heading-level styles to the h1-h6 heading styles.

For example, if you use a first-level heading style named "HeadText1" in FrameMaker, map it to "h1." Styles will map to the 'p' (paragraph) style by default.

19 Click **Next**.

20 If needed, map your character-level styles, like bold, to Flare styles.

Character-level styles will map to the "span" tag by default.

21 Click **Next**.

22 If needed, map your cross-reference (x-ref) styles to Flare styles.

Cross reference-styles will map to the MadCap|xref style by default.

23 Click **Import** and click **Accept**.

Importing a FrameMaker document into a project

You can import a FrameMaker document into an existing Flare project and divide it into multiple topics.

Shortcut	Tool Strip	Ribbon
Alt+P, I, F	Project > Import File > Add FrameMaker Import File	File > New

To import a FrameMaker document:

1 Select **File** > **New**.
—OR—
Right-click the **Imports** folder and select **Add FrameMaker Import File**.
The Add File dialog box appears.

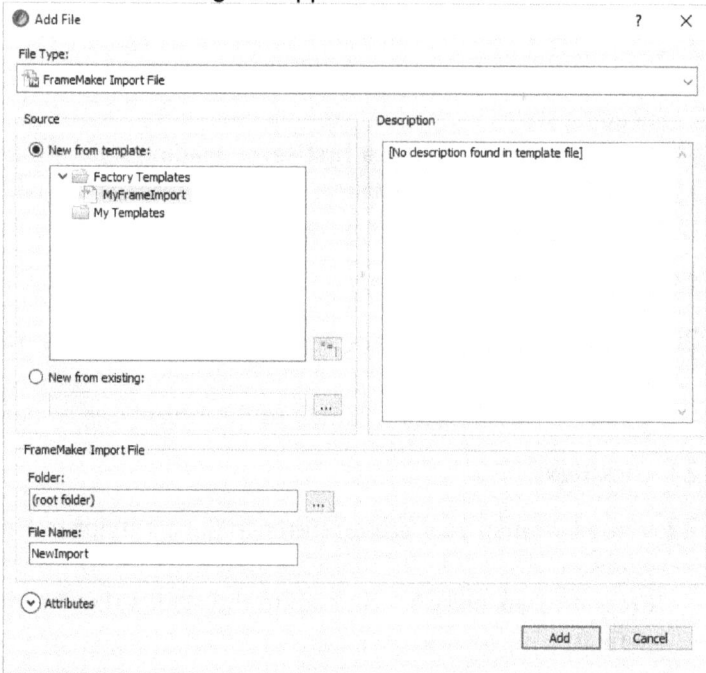

2 For **File Type**, select **FrameMaker Import File**.

3 Select a **Source** template.

4 Type a **File Name**.

5 Click **Add**.
The Frame Import Editor appears.

6 Click [+].
The Open dialog box appears.

7 Select a .fm or .book FrameMaker document and click **Open**. If needed, you can select more than one FrameMaker document.

8 If you plan to continue editing the file in FrameMaker, select **Link generated files to source files**.

◇ *This option allows you to link the imported topics to the FrameMaker document. When you re-import the document, Flare will replace the original topics with the new topics.*

Linked topics have a chain (⬡) icon after their file name when opened in the XML Editor.

9 Select the **New Topic Styles** tab.

10 Select a style or styles to use to create new topics.
For example, you can create a new topic for each Heading 1 in the FrameMaker document.

11 Select the **Options** tab.

12 If your images have callouts, select **Generate Images with Callouts**.
This option will create a MadCap Capture "props" file for the callouts, and you can edit the image and callouts in Capture.

13 If you have sized your images in FrameMaker, select **Preserve Image Size**.

14 If you want to automatically reimport the FrameMaker document(s) when you create a target, select **Auto-reimport before 'Generate Output.'**

15 If your document contains equations, select **Convert equations to MathML**.

16 Select the **Stylesheet** tab.

17 Select a stylesheet for the new topic(s).
If you select a stylesheet, Flare will apply the stylesheet to your topics and automatically "map" styles with matching names. If you do not select a stylesheet, Flare will create a stylesheet based on the formatting in your FrameMaker document.

18 Select the **Paragraph Styles** tab.

19 Map (or "match") your Word heading-level styles to the h1-h6 heading styles.

For example, if you use a first-level heading style named "HeadText1" in FrameMaker, map it to "h1." Styles will map to the "p" (paragraph) style by default.

20 Select the **Character Styles** tab.

21 If needed, map your FrameMaker character styles to Flare styles.

Character-level styles will map to the "span" tag by default.

22 Select the **Cross-Reference Styles** tab.

23 If needed, map your FrameMaker cross-reference (x-ref) styles to Flare styles.

Cross reference-styles will map to the MadCap|xref style by default.

24 Click **Import** in the toolbar.

The Accept Imported Documents dialog box appears.

25 Click **Accept**.

The imported topic or topics appear in the Content Explorer in a folder named after the FrameMaker import file you used.

'What happens to my... ?'

The following table explains how FrameMaker's features convert to Flare.

FrameMaker Feature	Converts?	Comments
Character styles	✓	Added to a stylesheet. By default, they are associated with the span style tag.
Conditional Tags	✓	Stored in a condition tag set that is named based on your import file.
Cross reference styles	✓	Added to a stylesheet and associated with the MadCap|xref style tag.
Equations	✓	Converted to MathML or images.

FrameMaker Feature	Converts?	Comments
Images	✓	Added to a folder in the Content Explorer named after your import file.
Index keywords	✓	Maintained and appear with a green background in your topics. You can hide them if needed.
Inline formatting	✓	Maintained. Can also be converted to styles.
Master pages		Converted to page layouts.
Named destinations	✓	Maintained in topics.
Paragraph styles	✓	Added to a stylesheet. By default, they are associated with the p (paragraph) style tag.
Table styles	✓	Converted to table stylesheets.
"TOC" document	✓	Converted to a Flare TOC named after your import file.
Variables	✓	Added to a variable set that is named after your import file.

Sample questions for this section

1 A Flare project file has the following extension:
   A) .prj
   B) .hhp
   C) .htm
   D) .flprj

2 Which folder contains your topics and images?
   A) Files
   B) Content
   C) Project
   D) Source

3 Does Flare support Unicode?
   A) Yes
   B) No

4 Which of the following files can be created based on a template?
   A) Topics
   B) Stylesheets
   C) Snippets
   D) All of the above

5 Why should you select "Link generated files to source files?"
   A) To automatically re-import your Word or FrameMaker documents when you build a target.
   B) To keep editing your content in Word or FrameMaker.
   C) To import your links and cross references.
   D) To import images.

6 Which of the following features is imported from RoboHelp but is not set up?
   A) Index
   B) TOC
   C) Variables
   D) Skins

7 Can you import FrameMaker .fm files and .book files into Flare?
  A) Yes
  B) No

8 Which of the following FrameMaker features cannot be imported?
  A) Styles
  B) Index keywords
  C) Template pages
  D) All of them can be imported.

Topics

This section covers:

- ☐ Structure bars
- ☐ Special characters
- ☐ QR codes
- ☐ Equations
- ☐ Code snippets
- ☐ Importing Word documents
- ☐ Importing Excel workbooks
- ☐ Importing DITA files
- ☐ Importing HTML files
- ☐ Importing Markdown files
- ☐ External resources
- ☐ Project linking
- ☐ Lists
- ☐ Tables
- ☐ Images
- ☐ Multimedia
- ☐ Slideshows

Creating topics

In Flare, your content is stored in topics. Each topic is a short (usually 1-4 printed pages) XHTML file that can contain formatted text, images, tables, lists, links, variables, snippets, and other types of content.

'What is XHTML?'

XHTML is a type (or "schema") of XML. XHTML files use HTML tags, but the tagging conforms to the strict rules of XML. For example, HTML does not require end tags for the <br>, <li>, or <img> tags. In XHTML, all tags must have end tags. So, a break is written in XHTML as <br />. Another difference is that HTML allows upper, lower, or mixed case tags: <BR>, <br>, or <Br>. In XHTML, you must use lowercase tags.

'Do I have to know XML to use Flare?'

Flare has a built-in WYSIWYG ("what-you-see-is-what-you-get") editor called the "XML Editor." You don't have to know anything about HTML, XHTML, or XML to use the XML Editor: Flare writes the code for you. If you *do* know how to write XHTML code, you can view the code and change it yourself.

Creating a topic

You can have as many topics in a project as you need. In fact, some Flare projects have over 100,000 topics.

Shortcut	Tool Strip	Ribbon
Ctrl+T	(Content Explorer)	File > New

To create a topic:

1 Select **File** > **New**.

   —OR—

   Click 🖹 in the Content Explorer toolbar.

   The Add File dialog box appears.

2 For **File Type**, select **Topic**.

3 Select a **Source** template.

4 Select a **Folder** to contain the new topic.

5 Type a **File Name** for the topic.

   You don't have to type the .htm extension. Flare will add it for you don't include it.

6 If you want to apply condition tag(s) or file tag(s) to the topic:

   ☐ Click **Attributes**.

   ☐ For **Condition Tags** and/or **File Tags**, click browse.

□ Select the condition tag(s) or file tag(s) and click **OK**.

7 Click **Add**.

The topic appears in the Content Explorer and opens in the XML Editor.

Viewing your topic titles

In Flare, your topics are listed by their filename in the Content Explorer. However, your users will view your topics by their topic titles in the index and search. Topic titles are also often used in link labels and in the TOC.

By default, your topic titles are automatically set to match the first heading in your topics. You can view and change your topic titles using the File List.

To view your topic titles:

1 Select **View** > **File List**.

2 Scroll to the right to the **Title** column.

TIP *You can click the Title column's heading and drag it to the left to make it easier to find.*

Opening a topic

Topics appear in the Content Explorer. When you double-click a topic, it opens as a new tab in the XML Editor. You can open as many topics as you need.

Shortcut	Tool Strip and Ribbon
Enter	📄 (Content Explorer toolbar)

To open a topic:

1 Select a topic in the Content Explorer.

2 Press **Enter**.
—OR—

Click 📑 in the Content Explorer toolbar.

The topic appears as a new tab in the XML Editor.

Opening a topic in the Text Editor

You can open a topic in the Text Editor to view and edit the XHTML code.

To open a topic in the Text Editor:

1 Open a topic.

2 Click the Text Editor tab at the bottom of the XML Editor window.

TIP> *If you click the Auto Complete button, Flare can automatically close your XHTML tags.*

Viewing a topic in a split-screen XML Editor and Text Editor

You can also view and edit topics in a split-screen Text Editor and XML Editor view.

To view a split-screen XML Editor and Text Editor:

1 Open a topic.

2 Click the ▲ Expand Pane button to the right of the XML Editor and Text Editor tabs.

TIP> *You can move the Text Editor to the right (rather than the bottom) by clicking the ▥ Vertical Split button to the right of the XML Editor and Text Editor tabs.*

Opening a topic in another editor TIP>

You can also open topics in other XHTML editors—if you know where to look!

To open a topic in another HTML editor:

1   Open the Content Explorer.

2   Right-click a topic and select **Open With** > *your XHTML editor*.
    The topic opens in the HTML editor you selected.

Opening two topics side by side

You can open multiple topics and switch between their tabs, or you can open two topics side by side to compare their content.

To open two topics side by side:

1   Open a topic.
    The topic opens as a new tab in the XML Editor.

2   Open another topic.
    The second topic opens as a new tab in the XML Editor.

3   Select **Window** > **Float**.
    The selected topic appears in a small floating window.

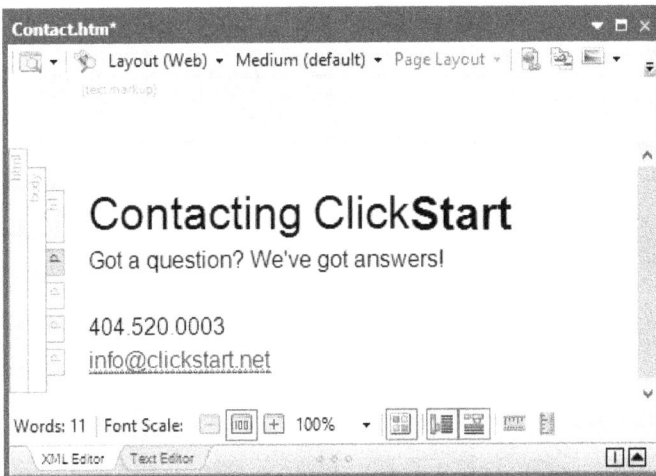

4   Click the floating window's title bar and drag the window.
    The positioning arrow appears.

5   Drag the window on top of one of the positioning squares.

6 Release your mouse button.

📑▷ *To move the window back to a tab, select* **Window** > **Float**, *and drag the window to the center positioning square.*

Rearranging open topic tabs

You can rearrange open topic tabs to make it easier to switch between topics.

To rearrange open topic tabs:

1 Open two or more topics.

2 Click an open topic's tab and drag it left or right.

3 Release the mouse button.
The topic's tab will move to its new location.

Using structure bars

In Flare, you can use the block and span structure bars to view the tagging behind your content. The block bar appears on the left side of the XML Editor, and the span bar appears on the top.

You can use the block bar to expand and collapse content within a block, such as a table or list, or all of the content between a heading and the next heading at the same level.

Icon	Description
▮▤	Show/hide block bar
▯▤	Show/hide span bar

To expand/collapse content:

1 Open a topic.

2 Position your cursor to the left of a heading or block element.
The triangular expand/collapse icon appears.

3 Click the expand/collapse icon.
The content expands/collapses.

TIP *You can click ⊞ or ⊟ in the XML Editor toolbar to expand or collapse all of the headings in a topic.*

Inserting a special character

You can insert special characters into your topics, such as ®, ™, ©, non-breaking spaces, and non-breaking hyphens. Common characters are available in the list when you select **Insert** > **Character** or click the ⓐ button in the toolbar. You can also add favorite characters to the list or set the quick character (inserted when you press **F11**).

Shortcut	Tool Strip	Ribbon
F11	ⓐ (XML Editor)	Insert > Character

To insert a special character:

1 Open a topic.

2 Position your cursor where you want to insert the special character.

3 Click the ⓐ button's down arrow.

4 Select a character.

To add a special character to the favorites group:

1 Select **Insert** > **Character**.
The Character dialog box appears.

2 Select a character.

3 Click ⭐.

4 Click **Close**.

To set the quick character:

1 Select **Insert** > **Character**.
The Character dialog box appears.

2 Select a character.

3 Click ⟦🔣⟧.

4 Click **Close**.

Inserting a QR code

A QR code is a type of barcode that can be read by a smart phone application. You can insert a QR code to display additional information, open a website, or send an email.

Shortcut	Tool Strip and Ribbon
Ctrl+Q	Insert > QR Code

To insert a QR code:

1 Open a topic.

2 Position your cursor where you want to insert the QR code.

3 Select **Insert** > **QR Code**.
The Insert QR Code dialog box appears.

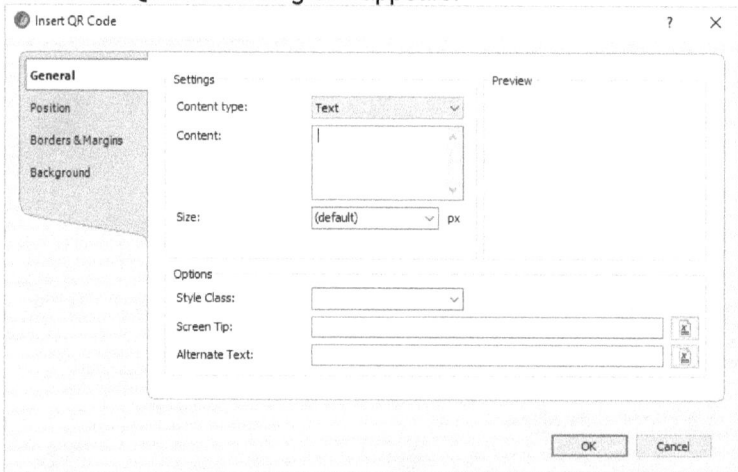

4 On the General tab, select a **Content Type**.

5 Type the **Content**.

6 Select a **Size**.

7 Type an **Alternate Text** description of the QR code.
For example, "QR code link to www.mywebsite.com." Alt text
is recommended by Section 508 of the U.S Government's
Rehabilitation Act and the W3C's Web Content Accessibility
Guidelines (WCAG).

8 Click **OK**.

Inserting an equation

You can use Flare's Equation Editor to insert an equation into a topic.
Or, you can create an equation in any equation editor (rather than
inside Flare) and paste it into the Equation editor.

◇ *If you insert an equation into a heading, the equation will not
appear in cross reference link text or in your table of contents.*

Shortcut	Tool Strip and Ribbon
Ctrl+E	Insert > Equation

To insert an equation:

1 Create the equation in an equation editor such as LaTeX.

TIP▶ *There are numerous free online equation editors if you
don't have one to use.*

2 Open a topic.

3 Position your cursor where you want to insert the equation.

4 Select **Insert > Equation**.
The Equation Editor appears.
◇ *The Equation Editor requires the Java Runtime
Environment (JRE) or OpenJDK. If prompted, download the
latest version from
www.madcapsoftware.com/downloads/java.aspx*

5 Paste the equation into the **Text Editor**.

6 Type an **Alternate Text** description of the equation. For example, alt text for "E=mc²" would be "E equals m c squared." Alt text is recommended by Section 508 of the U.S Government's Rehabilitation Act and the W3C's Web Content Accessibility Guidelines (WCAG).

7 Click **OK**.

Inserting a code snippet

You can use Flare's Code Editor to insert a code snippet into a topic. Code snippets can include captions and/or line numbers. Code snippets will automatically include syntax highlighting in online and print targets. Code snippets also automatically include a "Copy" button in HTML5 targets.

Shortcut	Tool Strip and Ribbon
Alt+N, C, P	Insert > Code Snippet

To insert a code snippet:

1 Open a topic.

2 Position your cursor where you want to insert the code snippet.

3 Select **Insert** > **Code Snippet**.
The Code Editor dialog box appears.

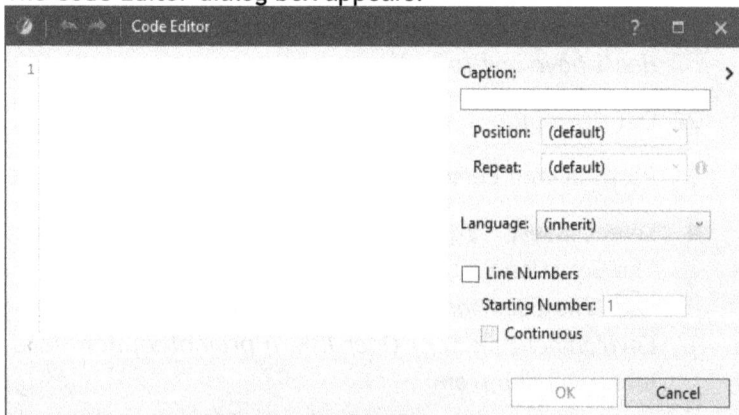

4 Type or paste the code snippet.

5 If you want to include a caption with your code snippet, type a **Caption** and select a **Position** and **Repeat** option.

6 Select a **Language**.

7 If you want to include line numbers, select **Line Numbers**.

8 Click **OK**.

Inserting an iframe

An iframe ("inline" frame) can be used to embed another webpage, video, or document into a topic, snippet, or template page.

Shortcut	Tool Strip	Ribbon
Alt+N, F, M	none	Insert > IFrame

To insert an iframe:

1 Open a file.

2 In the XML Editor, position your cursor where you want to insert the iframe.

3 Select **Insert > IFrame**.
The Insert IFrame dialog box appears.

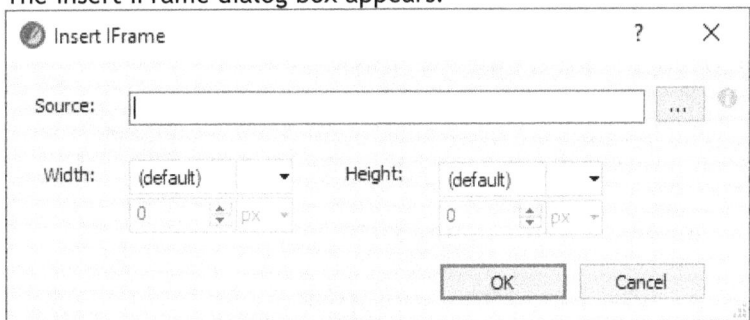

4 Type a **Source** URL.
—OR—
Click [...] and select a file.
Web-based URLs should use the https protocol.

5 Select or type a **Width.**

6 Select or type a **Height.**

7 Click **OK.**

Using smart quotes

You can use smart quotes instead of straight quotes in your topics. In English, smart quotes appear as " and " ("curly quotes"). If your content is set to a different language, Flare will use the language-appropriate quotes. To specify a language, see "Internationalization" on page 473.

Shortcut	Tool Strip	Ribbon
Alt+F, T	Tools > Options	File > Options

To use smart quotes:

1 Select **File > Options.**
The Options dialog box appears.

2 Select the **XML Editor** tab.

3 Select **Replace straight quotes with smart quotes.**

4 Click **OK.**

Finding an open topic

If Flare does not have enough room to display each topic's tab, the additional topics appear in a drop-down list.

To find an open topic in the XML Editor:

1 Click the ≡ down arrow on the right side of the XML Editor.

2 Select a topic in the drop-down list.

Closing all open topics

When you start using Flare, you will probably forget to close topics. If you have too many topics open, you can close all of them at once.

To close all open topics:

☐ Select **Window** > **Close All Documents.**
All of the open documents close.

To close all open topics except the current topic:

☐ Select **Window** > **Close All Documents Except This One.**
All of the open documents close except the current document.

Previewing a topic

You can preview a topic to see exactly how it will look in a target. Flare includes WYSIWYG ("what you see is what you get") editors, but they aren't always 100% accurate. That's normal for any HTML editor. If your formatting looks different in the preview, the preview is correct.

Shortcut	Tool Strip	Ribbon
Ctrl+W	View > Preview	View > Preview Window

To preview a topic:

1 Open a topic.

Click 🔍 in the XML Editor toolbar.

TIP► *By default, your topic will preview as it will appear in your primary target. Click the preview button's drop-down arrow to preview your topic in a different target.*

2 If you want to view a live preview as you edit a topic, you can drag the preview window to another screen or dock it on the right or left side of the Flare window. The preview will update when you open a new topic, save the topic, or click ⟳ in the Preview window's toolbar.

Deleting a topic

When you delete a topic, Flare moves the topic to the Windows recycle bin. If you need to temporarily remove a topic from your project, you can assign a condition tag to the topic and exclude it from your targets. For more information, see "Applying a tag to content" on page 315.

Shortcut	Tool Strip	Ribbon
Delete	✖ (Standard toolbar)	Home > Delete

To delete a topic:

1 Select the topic in the Content Explorer or File List.

2 Press **Delete**.
   The Delete confirmation dialog box appears.

3 Click **OK**.
   If anything links to the topic, the Link Update dialog box appears.

4 Click **Remove Links**.
   The topic is moved to the recycle bin.

Importing Word documents

When you import a Word document, you can divide the document into smaller topics based on styles that are used in your Word document. For example, you can create a new topic for each "Heading 1" in the document.

Flare can convert your Word template (.dot or .dotx file) into a stylesheet (.css file) so your formatting stays the same. Or, you can apply an existing stylesheet to reformat your imported topics to match your other topics.

◇ *Flare requires Word 2003 or later to import Word documents.*

Creating a new project based on a Word document

You can create a new project based on a Word document. If you want to import a Word document into an existing project, see "Importing a Word document" on page 86.

◇ *When you import a Word document, Flare saves your settings in an MS Word Import File (these files have a .flimp extension). You can reuse the import file when you re-import the Word document, or you can use it to import similar documents.*

Shortcut	Tool Strip	Ribbon
none	File > Import Project > MS Word Documents	File > New Project > MS Word Documents

To create a new project based on a Word document:

1   Select **File** > **New Project** > **MS Word Documents**.
    The Import Microsoft Word wizard appears.

Import Microsoft Word Wizard

Import Microsoft Word Documents...

Required

General

Optional

Styles

Advanced Options

Import into a new project:

Project name:

Sample Project

Output type:

HTML5

Project folder:

C:\Users\admin\Desktop

Add file   Remove file

MS Word Files   Status

Finish   Cancel

2 Type a **Project Name.**

3 Select an **Output Type.**
You can add additional output types to your project later.

4 Type or select a **Project Folder.**

5 Click **Add File.**

6 Select a Word document and click **Open.**
If needed, you can select more than one Word document.

7 Select the **Styles** tab.

8 Select a stylesheet for the new topic(s).
If you select a stylesheet, Flare will apply the stylesheet to your topics and automatically "map" styles with matching names. If you do not select a stylesheet, Flare will create a stylesheet based on the formatting in your Word document(s).

9 Select a style or styles to use to create new topics.
For example, you can create a new topic for each Heading 1 in the Word document(s).

10 Map (or "match") your Word paragraph-level styles to your stylesheet's (.css) styles.
For example, if you use a style named "HeadText1" in Word, map it to "h1."

11 If you want to create new topics based on a Word heading style, check the checkbox in the style's **Start new topic on** column.

12 Map any character-level styles, like bold, to your stylesheet's styles.
Character-level styles will map to the "span" tag by default.

13 Select the **Advanced Options** tab.

14 If you want to create a new stylesheet based on the formatting in the imported document(s), select **Create new stylesheet**.

15 Select how you want to import inline formatting.

16 If you want to set the first row of every table as a header row, select **Set first row of each table as a header row**. Setting the first row as a header row will allow you to provide different formatting to the first row of each table.

17 Select how you want to import table styles.

18 If you want to convert your bullets and numbers to standard bullet icons and numbers, select **Use standard list type**.

   ◇ *If you use custom bullet icons or number formats, you can clear the option to keep your custom lists. However, your custom lists will import with inline formatting.*

19 If your Word document(s) uses different headers and/or footers for different sections, select **Create a page layout for each section header/footer**.

20 Select whether you want to preserve or ignore page breaks.

21 If you plan to continue editing the file(s) in Word, select the **Link generated files to source files** and **Auto-reimport before generate output** options.
   ◇ *These options allow you to link the imported topics to the Word document(s). When you re-import the document(s), Flare will replace the original topics with the new topics.*

   Linked topics have a chain (⬚) icon after their file name when opened in the XML Editor.

22 Click **Finish.**

Importing a Word document

You can import a Word document into an existing Flare project and divide it into multiple topics.

Shortcut	Tool Strip	Ribbon
Alt+P, I, W	Project > Import File > Add MS Word Import File	File > New

To import a Word document:

1 Select **File** > **New.**
—OR—
Right-click the **Imports** folder and select **Add MS Word Import File.**
The Add File dialog box appears.

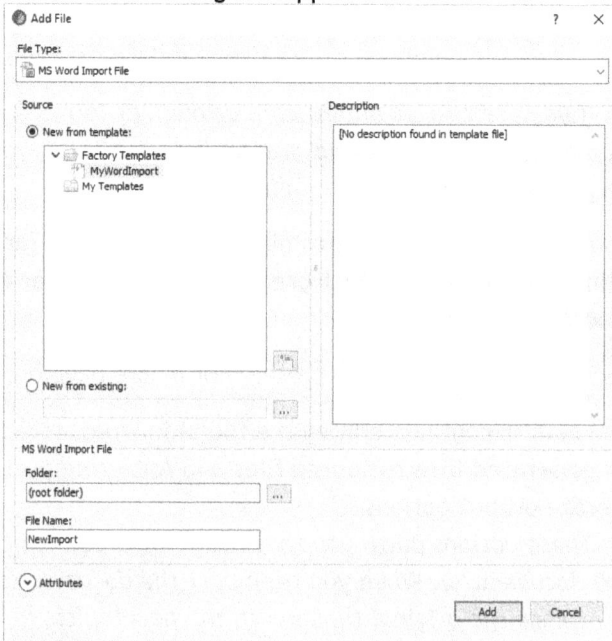

2 For **File Type**, select **MS Word Import File.**

3 Select a **Source** template.

4 Type a **File Name**.

5 Click **Add**.
The MS Word Import Editor appears.

6 For **Import to folder**, type or select a folder for the imported topics.

7 Click **Add File**.
The Open dialog box appears.

8 Select a Word document and click **Open**.
If needed, you can select more than one Word document.

9 Select the **Styles** tab.

10 Select a stylesheet for the new topic(s).
If you select a stylesheet, Flare will apply the stylesheet to your topics and automatically "map" styles with matching names. If you do not select a stylesheet, Flare will create a stylesheet based on the formatting in your Word document(s).

11 Map (or "match") your Word paragraph-level styles to your stylesheet's (.css) styles.
For example, if you use a style named "HeadText1" in Word, map it to "h1." Styles will map to the "p" (paragraph) style by default.

12 If you want to create new topics based on a Word heading style, check the checkbox in the style's **Start new topic on** column.

13 Map any character-level styles, like bold, to your stylesheet's styles.
Character-level styles will map to the "span" tag by default.

14 Select the **Advanced Options** tab.

15 If you want to create a new stylesheet based on the formatting in the imported document(s), select **Create new stylesheet**.

16 Select how you want to import inline formatting.

17 If you want to set the first row of every table as a header row, select **Set first row of each table as a header row**. Setting the first row as a header row will allow you to provide different formatting to the first row of each table.

18 Select how you want to import table styles.

19 If you want to convert your bullets and numbers to standard bullet icons and numbers, select **Use standard list type**.

◇ *If you use custom bullet icons or number formats, you can clear the option to keep your custom lists. However, your custom lists will import with inline formatting.*

20 If your Word document(s) uses different headers and/or footers for different sections, select **Create a page layout for each section header/footer**.

21 Select whether you want to preserve or ignore page breaks.

22 If you plan to continue editing the file(s) in Word, select the **Link generated files to source files** and **Auto-reimport before generate output** options.

◇ *These options allow you to link the imported topics to the Word document. When you re-import the document(s), Flare will replace the original topics with the new topics.*

Linked topics have a chain (⬡) icon after their file name when opened in the XML Editor.

23 Click **Import** in the toolbar.
The Accept Imported Documents dialog box appears.

24 Click **Accept**.
The imported topic or topics appear in the Content Explorer in a folder named after the Word import file you used.

Importing Excel workbooks

When you import an Excel workbook, you can divide the workbook into topics based on worksheets in your workbook. You can even select which worksheets are imported.

Flare can maintain formatting from Excel, or you can apply a Flare table style to the imported content.

◇ *Flare requires Excel 2010 to maintain formatting from Excel's built-in "factory" styles.*

Creating a new project based on an Excel workbook

You can create a new project based on an Excel workbook. If you want to import an Excel workbook into an existing project, see "Importing an Excel workbook" on page 86.

◇ *When you import an Excel workbook, Flare saves your settings in an MS Excel Import File (these files have a .flimpxls extension). You can reuse the import file when you re-import the Excel workbook, or you can use it to import similar workbooks.*

Shortcut	Tool Strip	Ribbon
none	File > Import Project > MS Excel Workbooks	File > New Project > MS Excel Workbooks

To create a new project based on an Excel workbook:

1 Select **File** > **New Project** > **MS Excel Workbooks**.
The Import Microsoft Excel wizard appears.

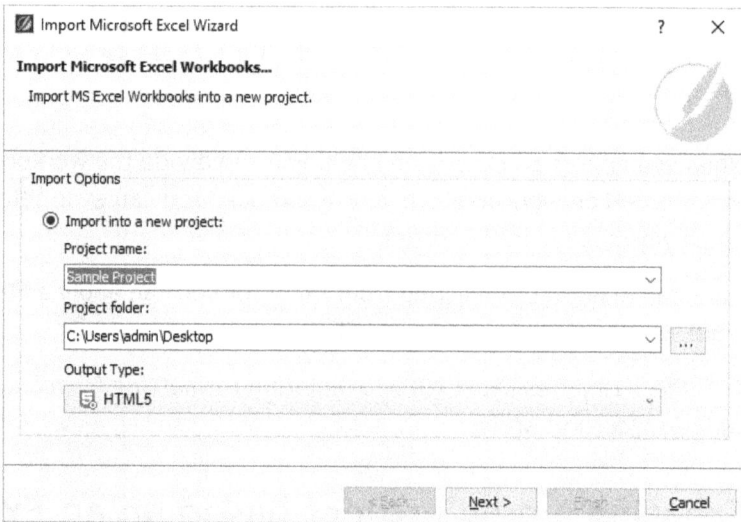

2 Type a **Project Name**.

3 Type or select a **Project Folder**.

4 Select an **Output Type**.
You can add additional output types to your project later.

5 Click **Next**.

6 Click 🔲.

7 Select an Excel workbook and click **Open**.
If needed, you can select more than one Excel workbook.

8 If you plan to continue editing the content in Excel, select **Link generated files to source files**.

✎ *This option allows you to link the imported topics to the Excel workbook. When you re-import the workbook, Flare will replace the original topics with the new topics.*

Linked topics have a chain ( ) icon after their file name when opened in the XML Editor.

9 Click **Next**.

10 Select whether you want to import the workbook(s) as topics or snippets.

11 If you are importing content as snippets, select whether you want to **Include tab title as headings** and/or **Create a topic with snippets**.

12 If you want to organize your content into a separate folder for each workbook, select **Organize worksheets into different folders per workbook**. If you do not select this option, all of your imported content will be organized into one folder.

13 If you want to import hidden content, select **Import hidden rows/columns**.

14 If you want to set the first row a header row, select **Use first row as column header**.
Setting the first row as a column header will allow you to provide different formatting to the first row of each table.

15 If your Excel workbook(s) contains equations, select, select **Import equations based on settings**.

16 If your Excel workbook(s) contain charts, Select **Import charts as** and select an image format.

17 If you want to create new topics/snippets based on a maximum number of rows, select **Split topics or snippets by maximum rows** and select a number of rows.

18 If you want to automatically reimport the Excel workbook(s) when you create a target, select **Auto-reimport before 'generate output.'**

19 Click **Next**.

20 Select whether you want to **Preserve MS Excel styles**.
If you preserve Excel styles, your imported content will use the same formatting in Flare as it did in Excel. If you do not preserve styles, you can apply a Flare table style to your imported content.

21 Click **Next**.

22 Select the worksheet(s) you want to import.

23 Click **Finish**.

Importing an Excel workbook

You can import an Excel workbook into an existing Flare project and divide the worksheets into multiple topics.

Shortcut	Tool Strip	Ribbon
Alt+P, I, X	Project > Import File > Add MS Excel Import File	File > New

To import an Excel workbook:

1 Select **File** > **New**.

   —OR—

   Right-click the **Imports** folder and select **Add MS Excel Import File**.

   The Add File dialog box appears.

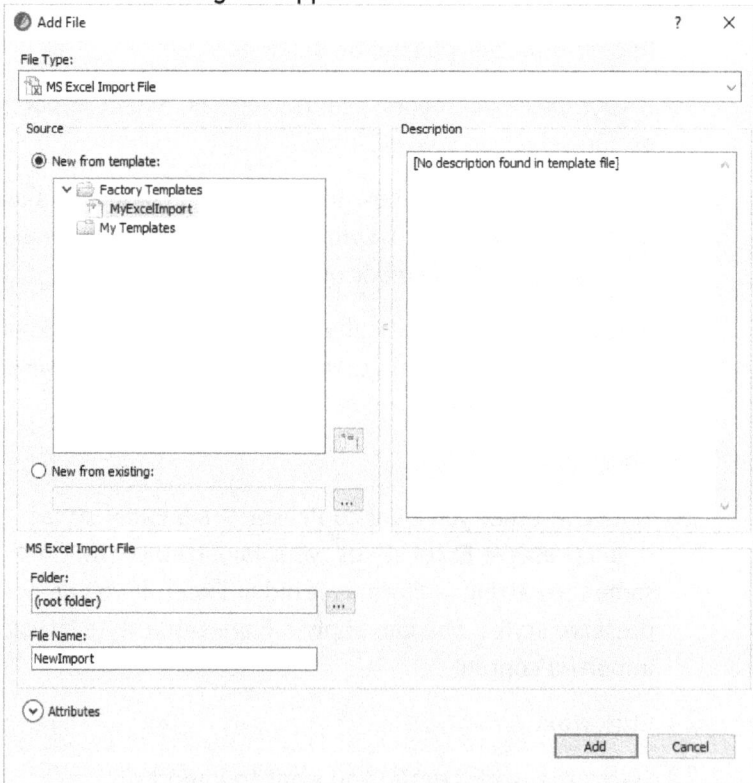

2 For **File Type**, select **MS Excel Import File**.

3 Select a **Source** template.

4 Type a **File Name**.

5 Click **Add**.
The MS Excel Import Editor appears.

6 Click ⊞.
The Open dialog box appears.

7 Select an Excel workbook and click **Open**.
If needed, you can select more than one Excel workbook.

8 If you plan to continue editing the content in Excel, select **Link generated files to source files**.
◇ *This option allows you to link the imported topics to the Word document. When you re-import the document, Flare will replace the original topics with the new topics.*

Linked topics have a chain (🔗) icon after their file name when opened in the XML Editor.

9 Select the **Options** tab.

10 Select whether you want to import the workbook(s) as topics or snippets.

11 If you are importing content as snippets, select whether you want to **Include tab title as headings** and/or **Create a topic with snippets**.

12 If you want to organize your content into a separate folder for each workbook, select **Organize worksheets into different folders per workbook**. If you do not select this option, all of your imported content will be organized into one folder.

13 If you want to import hidden content, select **Import hidden rows/columns**.

14 If you want to set the first row a header row, select **Use first row as column header**.
Setting the first row as a column header will allow you to provide different formatting to the first row of each table.

15 If your Excel workbook(s) contains equations, select, select **Import equations based on settings**.

16 If your Excel workbook(s) contain charts, Select **Import charts as** and select an image format.

17 If you want to create new topics/snippets based on a maximum number of rows, select **Split topics or snippets by maximum rows** and select a number of rows.

18 If you want to automatically reimport the Excel workbook(s) when you create a target, select **Auto-reimport before 'generate output.'**

19 Select the **Styles** tab.

20 Select whether you want to **Preserve MS Excel styles**. If you preserve Excel styles, your imported content will use the same formatting in Flare as it did in Excel. If you do not preserve styles, you can apply a Flare table style to your imported content.

21 Select the **Filters** tab.

22 Select the worksheet(s) you want to import.

23 Click **Import** in the toolbar. The Accept Imported Documents dialog box appears.

24 Click **Accept**. The imported topic(s) or snippet(s) appear in the Content Explorer.

Importing DITA files

You can create a new project based on DITA files, or you can import DITA files into an existing Flare project. If you link the imported topics to their source files, you can edit the DITA files and re-import them into your project.

Creating a new project based on a DITA document set

You can create a new project based on a .dita or .ditamap file. If you want to import a DITA document or DITA map into an existing project, see "Importing a DITA document" on page 97.

When you import a DITA file or map, Flare saves your settings in a DITA Import File (these files have a .flimpdita extension). You can reuse these settings when you re-import the DITA file or import similar files.

Shortcut	Tool Strip	Ribbon
none	Project > Import Project > DITA Document Set	File > New Project > DITA Document Set

To create a new project based on a DITA document set:

1 Select **File** > **New Project** > **DITA Document Set**.
   The Import DITA wizard appears.

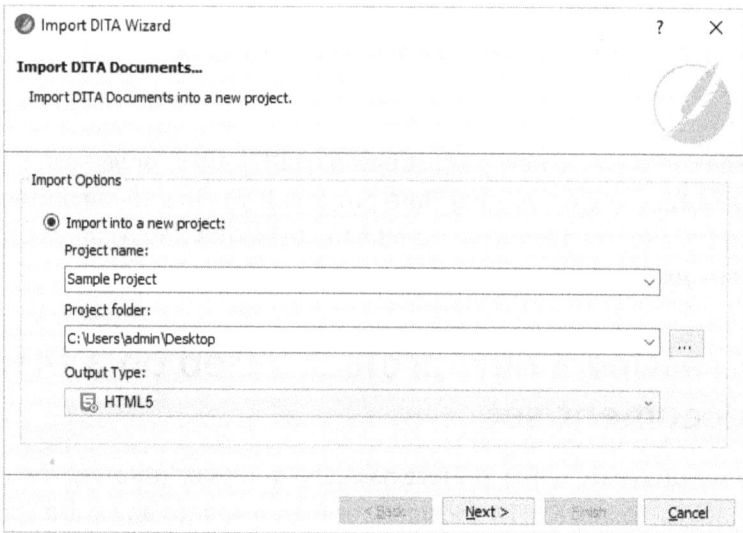

2   Type a **Project Name**.

3   Type or select a **Project Folder**.

4   Select an **Output Type**.
    You can add additional output types to your project later.

5   Click **Next**.

6   Click [icon].

7   Select a .dita or .ditamap file and click **Open**.
    If needed, you can select more than one DITA file.

8   If you plan to continue editing the original DITA files, select
    Link generated files to source files.
    ◇ *This option allows you to link the imported topics to the*
    source DITA files. When you re-import the file, Flare will
    replace the original topics with the new topics.

    Linked topics have a chain (⊗) icon after their file name when
    opened in the XML Editor.

9   Click **Next**.

10 Select **Import all content files to one folder** if you want to
    import all of the DITA files into one folder.

11 Select '**Auto-reimport before Generate Output**' if you want to automatically re-import the DITA document(s) when you generate a target.

12 Select **Preserve ID attributes for elements** if you plan to build a DITA target from your project.

13 Click **Next**.

14 Click **Conversion Styles** if you want to change the formatting of your topics.
Flare will create style classes for your DITA tags. You can modify each tag's formatting later in your stylesheet.

15 Select a stylesheet for the new topic(s).
If you select a stylesheet, Flare will apply the stylesheet to your topics.

16 Click **Finish**.

Importing a DITA file

You can import a .dita or .ditamap file into an existing Flare project.

Shortcut	Tool Strip	Ribbon
Alt+P, I, D	Project > Import File > Add DITA Import File	File > New

To import a DITA file or DITA map:

1 Select **File** > **New**.
—OR—
Right-click the **Imports** folder and select **Add DITA Import File**.
The Add File dialog box appears.

```
Add File                                           ?   ×
File Type:
  DITA Import File                                          ∨
┌─Source──────────────────────┬─Description──────────────┐
│ ⦿ New from template:         │ [No description found in template file]  ∧│
│   ∨ 📁 Factory Templates      │                          │
│       📄 MyDITAImport          │                          │
│       📁 My Templates          │                          │
│                              │                          │
│                              │                          │
│                              │                          │
│                              │                          │
│                       [📋]    │                          │
│ ○ New from existing:         │                          │
│   [            ]  [...]       │                          ∨│
└─────────────────────────────┴──────────────────────────┘
┌─DITA Import File───────────────────────────────────────┐
│ Folder:                                                 │
│ (root folder)                            [...]          │
│ File Name:                                              │
│ NewImport                                               │
└────────────────────────────────────────────────────────┘
  (∨) Attributes
                                    [  Add  ]  [ Cancel ]
```

2   For **File Type**, select **DITA Document Set**.

3   Select a **Source** template.

4   Type a **File Name**.

5   Click **Add**.
    The DITA Import Editor appears.

6   Click ⊕.

7   Select a .dita or .ditamap file and click **Open**.
    If needed, you can select more than one DITA file.

8   If you plan to continue editing the original DITA files, select
    Link generated files to source files.
    ◇ *This option allows you to link the imported topics to the*
    source DITA files. When you re-import the file, Flare will
    replace the original topics with the new topics.

Linked topics have a chain (⚭) icon after their file name when opened in the XML Editor.

9 Select the **Options** tab.

10 Select **Import all content files to one folder** if you want to import all of the DITA files into one folder.

11 Select **'Auto-reimport before Generate Output'** if you want to automatically re-import the DITA file(s) when you generate a target.

12 Select **Preserve ID attributes for elements** if you plan to build a DITA target from your project.

13 Select the **Stylesheet** tab.

14 Click **Conversion Styles** if you want to change the formatting of your topics.
Flare will create style classes for your DITA tags. You can modify each tag's formatting later in your stylesheet.

15 Select a stylesheet for the new topic(s).
If you select a stylesheet, Flare will apply the stylesheet to your topics.

16 Click **Import** in the toolbar.
The Accept Imported Documents dialog box appears.

17 Click **Accept**.
The imported topic or topics appear in the Content Explorer in a folder named after the DITA import file you used.

Importing HTML and XHTML files

You can import HTML and XHTML files into a project. If you import an HTML file, Flare will convert it to XHTML.

TIP *If you need to import an Acrobat PDF file, save your PDF file as HTML and import the HTML file.*

To import an HTML or XHTML file:

1 Select **File** > **New**.
—OR—
Right-click the **Imports** folder and select **Add HTML Import File**.
The Add File dialog box appears.

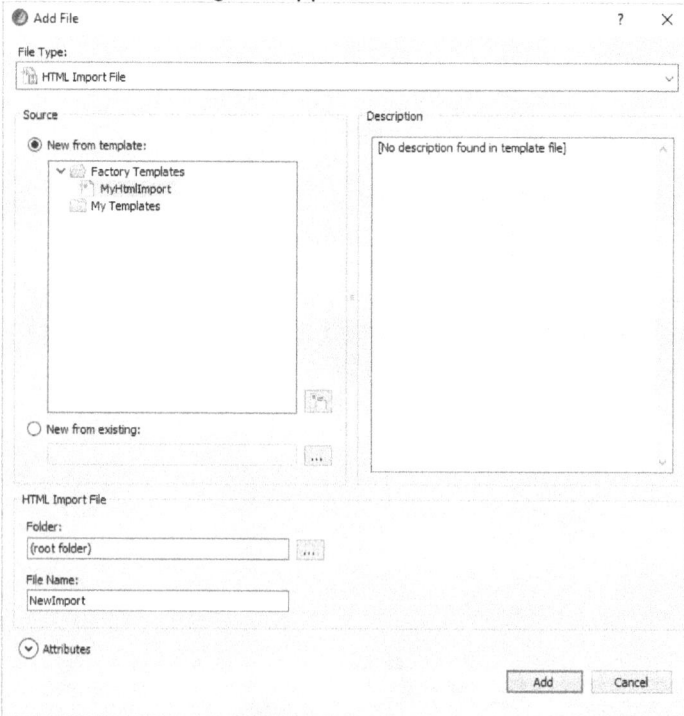

2 For **File Type**, select **HTML Document Set**.

3 Select a **Source** template.

4 Type a **File Name**.

5 Click **Add**.
The HTML Import Editor appears.

6 Click .

7 Select a .htm, .html, or .xhtml file and click **Open**.
If needed, you can select more than one file.

8 If you plan to continue editing the original file(s), select **Link generated files to source files**.

9 Select the **Options** tab.

10 Select a folder for the imported files.

11 Select **Import linked HTML files** if you want to also import any other files that the selected files link to.
For example, if you import FileA and it links to FileB, this option will also import FileB.

12 Select **Import resources** if you also want to import any files that are used by the selected file(s).
For example, images, stylesheets, or script files used in a topic.

13 If you want to re-import the files when you generate a target, select **Auto-reimport before 'Generate Output.'**

14 Click **Import** in the toolbar.
The Accept Imported Documents dialog box appears.

15 Click **Accept**.
The imported topic or topics appear in the Content Explorer in the selected folder.

Importing Markdown files

You can import Markdown files into a project. If you import a Markdown file, Flare will convert it to XHTML.

To import a Markdown file:

1 Select **File** > **New**.

—OR—

Right-click the **Imports** folder and select **Add Markdown Import File**.

The Add File dialog box appears.

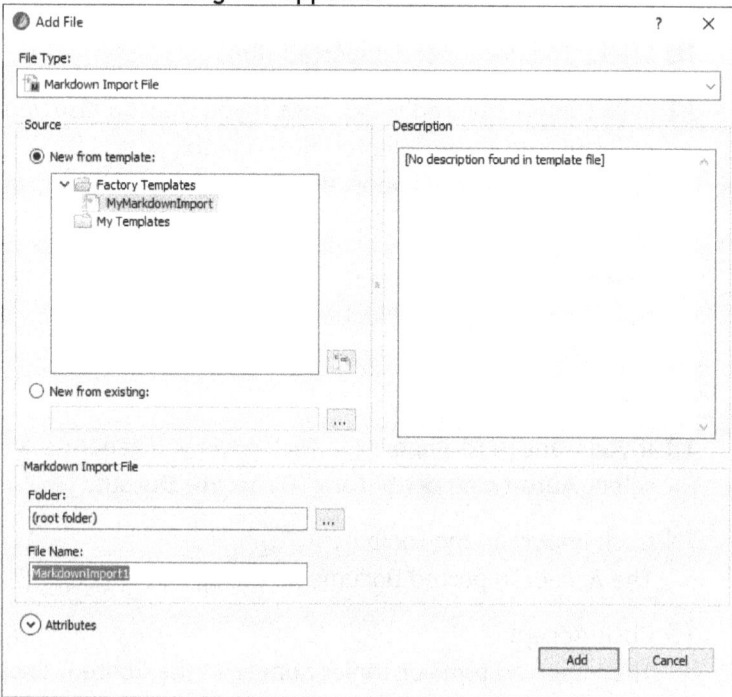

2 For **File Type**, select **Markdown Import File**.

3 Select a **Source** template.

4 Type a **File Name**.

5 Click **Add**.

The Markdown Import Editor appears.

6 For **Import to folder**, type or select a folder for the imported topics.

7 Select a .md file and click **Open**.

If needed, you can select more than one file.

8 Select the **Styles** tab.

9 Select a stylesheet for the new topic(s).

If you select a stylesheet, Flare will apply the stylesheet to your topics and automatically "map" styles with matching names. If you do not select a stylesheet, Flare will create a stylesheet based on the formatting in your Markdown file(s).

10 Map (or "match") your Markdown styles to your stylesheet's (.css) styles.

11 If you want to create new topics based on a heading level, check the checkbox in the style's Start new topic on column.

12 Select the **Advanced Options** tab.

13 If you want to create a new stylesheet based on the formatting in the imported document(s), select **Create new stylesheet**.

14 If you want to add paragraph tags to list items, select **Convert all simple lists to paragraph lists**.

15 If you want to add paragraph tags to list items, select **Convert all definition lists to paragraph lists**.

16 Select **Resources** if you also want to import any files that are used by the selected file(s).

For example, images, stylesheets, or script files used in a file.

17 Select how you want to import table styles.

18 If you plan to continue editing the Markdown file(s), select **Link generated files to source files**.

◇ *This option allows you to link the imported topics to the Markdown file(s). When you re-import the file(s), Flare will replace the original topics with the new topics.*

Linked topics have a chain (⊗) icon after their file name when opened in the XML Editor.

19 Click **Import** in the toolbar.

The Accept Imported Documents dialog box appears.

20 Click **Accept**.

The imported topic or topics appear in the Content Explorer in the selected folder.

Importing external resources

You can import files from other locations, such as a network drive, and use them in your project. If you synchronize the files, you can update the source or Flare version when either is modified.

TIP *If you use SharePoint, you can add files from SharePoint to your project as external resources. Your SharePoint site will appear at the bottom of the list when you select the external resource folder.*

Shortcut	Tool Strip and Ribbon
Alt+V, E	View > External Resources

To import an external resource:

1 Select **Project** > **External Resources**.

2 Click ⬚.
   The Select Folder dialog box appears.

3 Select the folder that contains the external resource.

4 Click **OK**.
   The selected folder appears in the External Resources pane.

5 Select the file(s) you want to import.

6 Click ⬚.
   The Select Project Path dialog box appears.

7 Select a folder and click **OK**.
   The Copy to Project dialog box appears.

8 If you want to be able to update the file when the source file is modified, select **Keep file(s) synchronized (create mapping)**.

9 Click **OK**.
   The file is added to your project. If it is synchronized, it will have a ⬚ icon.

To synchronize external resources:

1 Select **View** > **External Resources**.

2 Click 🗐.
The Synchronize Files dialog box appears.

3 Select **Synchronize Files**.

4 Click **Synchronize**.

5 Click **OK**.

'What happens if I move or rename the external resource?'

If you move, rename, or delete an external resource, Flare will not be able to find it when you click **Synchronize Files**. To fix it, highlight the file in the list, click **Manage Mappings**, and find the file again.

Importing content from Flare projects

You can link projects to reuse files in multiple projects. For example, you can create a template project that contains your stylesheet, page layout, template page, and a "Contact Us" topic. When you create a new project, you can link the new project to the template project. If you update the files in the source project, you can re-import them into the shared project.

Flare project import files are stored in the Imports folder in the Project Organizer.

Shortcut	Tool Strip	Ribbon
Alt+P, I, P	Project > Import File > Add Flare Project Import File	File > New

To import content from another Flare project:

1 Select **File** > **New**.
  —OR—
  Right-click the **Imports** folder and select **Add Flare Import File**.
  The Add File dialog box appears.

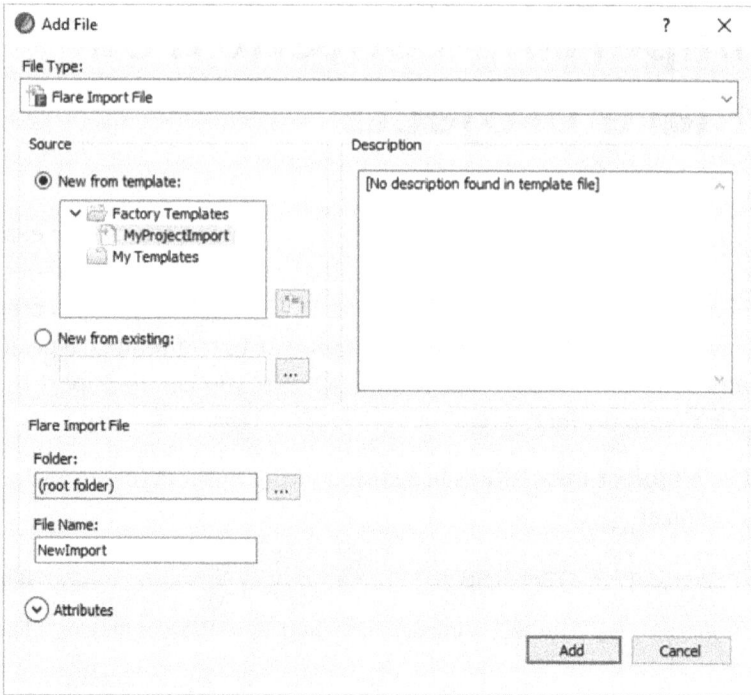

Add File dialog showing:

File Type:
Flare Import File

Source
● New from template:
  ∨ 📁 Factory Templates
      📄 MyProjectImport
  📁 My Templates

○ New from existing:

Description
[No description found in template file]

Flare Import File
Folder:
(root folder)

File Name:
NewImport

∨ Attributes

[Add] [Cancel]

2   For **File Type**, select **Flare Import File**.

3   Select a **Source** template.

4   Type a **File Name**.

5   Click **Add**.

6   Click **OK**.
    The Project Import Editor appears.

Project Import Editor | Import...

Source Project

Imported Files

Removed Links

Source Project:

☑ Auto-reimport before "Generate Output"   [Browse...]

☑ Delete stale files   [Open]

☐ Delete unreferenced files

Include Files:

Topic Files (*.htm; *.html)

☐ Auto-include linked files   [Edit...]

Exclude Files:

No Files

[Edit...]

Note: These fields let you select file types (e.g., topic files, snippet files). To import or exclude specific files: (1) save your changes in this editor; (2) click the Import or Re-Import button above; and (3) in the Accept Imported Documents dialog, make sure a check mark appears next to the files you want to include.

Import Conditions:

☐ Auto-Exclude Non-Tagged Files   [Edit...]

7 Click **Browse** and select a Flare project (.flprj) file.

8 If you want to re-import the shared files when you generate a target, select **Auto-reimport before 'Generate Output.'**

9 If you want to delete files from your project if they are deleted from the source project, select **Delete Stale Files.**

10 If you want to be able to select which files are reimported later, select **Delete Unreferenced Files.**

11 For **Include Files**, select the files or file types to be imported.

12 Select **Auto-include linked files** if you also want to import any files that are used by the selected files (for example, images, stylesheets, or script files used in a topic).

13 For **Exclude Files**, select the file types to not be imported.

14 Click **Import.**

TIP *You can include or exclude multiple files or file types. Here are some examples:*

Entry	Description
overview.htm	only the overview.htm topic
*.css	all .css files (stylesheets) in the project
*.css; overview.htm	all .css files and the overview.htm topic
MyCompany	all files that include "MyCompany" in their filename
.fl	all Flare-specific files (including TOCs, snippets, page layouts, template pages, and variables)

Lists

You can create bulleted lists to help users scan groups of items, or you can create numbered lists to provide step-by-step instructions. Flare provides the following XHTML list types:

Type	Example
Bulleted list	• Item • Item
Circle bulleted list	○ Item ○ Item
Square bulleted list	▪ Item ▪ Item
Numbered list	1. Item 2. Item
Lower-alpha numbered list	a. Item b. Item
Upper-alpha numbered list	A. Item B. Item
Lower-roman numbered list	i. Item ii. Item
Upper-roman numbered list	I. Item II. Item
Definition list	Term Description

You can modify the list styles to change the bullet icon or format the bullets or numbers.

Flare provides one toolbar button for both bulleted and numbered lists.

Creating a list

You can select the list type when you create the list.

Shortcut	Tool Strip	Ribbon
Alt+H, U, N	Format > List	☰ ▾ (Home ribbon)

To create a bulleted or numbered list:

1 Open a topic.

2 Position your cursor where you want to create the list.
   —OR—
   Highlight content that you want to format as a list.

3 Click the ☰ ▾ button's down arrow.

4 Select a list type.

5 If you are creating a new list, type the list items.

Sorting a list

You can use the tag bar to sort a list.

To sort a list:

1 Select the list.

2 If you are not viewing the tag bar, click ☰ in the XML Editor's lower toolbar.

3 Right-click the **ol** (numbered list) or **ul** (bulleted list) tag in the tag bar.

4 Select **Sort List** to sort the list.
   —OR—
   Select **Reverse List** to sort the list in reverse order.

Continuing lists numbers

If you have two lists, the second list probably restarts with "1." You can continue the numbering in the second list, if needed.

To continue list numbers:

- Right-click the second list's ol tag and select **Continue Sequence**.

Tables

You can create tables to organize content and to help users quickly find information. For example, this guide uses tables to present keyboard shortcuts, tool strip buttons, and ribbon commands.

Tables can contain any type of content, including images and lists, and they can be formatted with background shading and borders. You can have as many table rows or columns as you need, and Flare makes it easy to move, add, and delete columns and rows.

Creating a table

Shortcut	Tool Strip	Ribbon
Alt+B, T	(Format toolbar)	Insert > Table

To create a table:

1 Open a topic.

2 Position the cursor where you want to create the table.

3 Select **Insert** > **Table**.
   The Insert Table dialog box appears.

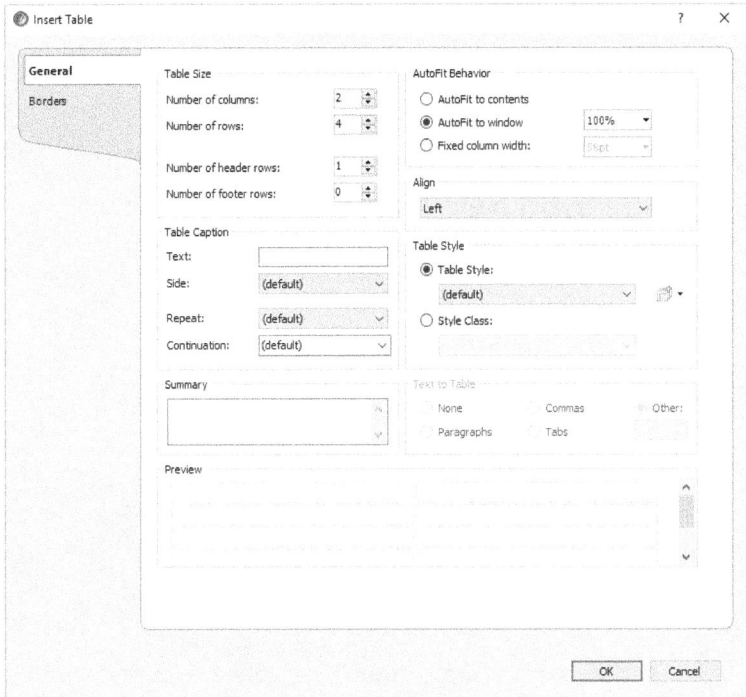

4 Type or select a **Number of Columns.**

5 Type or select a **Number of Rows.**

6 If needed, type or select a number of header and/or footer rows.

7 If needed, type a **Table Caption** and select a caption location.

8 If needed, type a table **Summary.**
Table captions and summaries are recommended by Section 508 of the U.S Government's Rehabilitation Act and the W3C's Web Content Accessibility Guidelines (WCAG).

9 Select a column width.

□ **AutoFit to Contents** — each column's width is based on the amount of content it contains.

□ **AutoFit to Window** — the columns are equally-sized to fit the size of the window.

☐ **Fixed Column Width** — each column is set to a specified width.

10 Click **OK**.

Converting text to a table

You can convert text to a table based on paragraphs, comma, or another character.

Shortcut	Tool Strip	Ribbon
none	Table > Insert > Table	Insert > Table

To convert text to a table:

1 Highlight the text.

2 Select **Insert** > **Table**.

3 In the **Text to Table** group box, select a conversion option.

4 Click **OK**.

Converting a table to text

You can also convert a table to text.

Shortcut	Tool Strip and Ribbon
Alt+B, V	Table > Convert to Text

To convert text to a table:

1 Select the table.

2 Select **Table** > **Convert to Text**.

Sorting table rows

You can sort a table in alphabetical or reverse alphabetical order based on the text in any of the table's columns. If you use Flare's advanced sorting options, you can sort by multiple rows/columns immediately or when you build a target.

Shortcut	Tool Strip and Ribbon
Alt+B, S, R	Table > Sort Rows

To sort table rows:

1 Click inside the column you want to use for sorting.

2 Select **Table** > **Sort Rows** > **Ascending** or **Descending**.

To use advanced sorting options:

1 Click inside the table.

2 Select **Table** > **Sort Rows** > **Advanced Sorting Options**. The Advanced Sorting Options dialog box appears.

3 Click 📄.

4 In the **Column** cell, click the down arrow and select a column for sorting.

5 In the **Direction** cell, select **ascending** or **descending**.

6 If you want to add another row to specify another sorting level:

   ☐ Click 📄.

   ☐ Select a **Column**.

   ☐ Select a **Direction**.

7 If you want to automatically update the table sorting when you build a target, select **Apply at compile time**.

8 If you want to apply the sorting rules immediately, click **Apply now**.

9 Click **OK**.

Rearranging table rows or columns

You can rearrange table rows or columns by dragging and dropping them in the structure bars.

To rearrange table rows or columns:

1   Select a row in the block bar or a column in the span bar.

2   Drag the row up/down or the column left/right.
    A blue arrow will appear.

3   Release the mouse button to move the row or column.

Inserting table rows or columns

When you insert rows or columns, you can specify whether they're inserted above or below the current row or column. You can also highlight multiple rows or columns to insert multiple rows or columns at the same time.

To insert table rows or columns:

1   Click inside a table. If you want to insert multiple rows or columns, highlight the number of rows or columns you want to insert.

2   Right-click and select **Insert**.

3   Specify columns or rows and above or below.

Creating a table style

In Flare, you can create table styles to format your tables. For example, you can create a table style named "noBorders" to create tables without borders and another named "greenHeading" to create tables with a green background for headings.

Table styles are stored in table stylesheets with a .css extension.

Shortcut	Tool Strip	Ribbon
Alt+B, N	📄 (Content Explorer)	File > New

To create a table style:

1 Select **File** > **New**.
The Add File dialog box appears.

2 For **File Type**, select **Table Style**.

3 Select a **Source** template.

4 Select a **Folder**.
By default, table stylesheets are stored in the Resources/
TableStyles folder.

5 Type a **File Name** for the table stylesheet.
Table stylesheets have a .css extension. If you don't type the
extension, Flare will add it for you.

6 Click **Add**.
The table stylesheet appears in the Content Explorer and opens
in the TableStyle Editor.

To modify a table style:

1 Open a table stylesheet.
The TableStyle Editor appears.

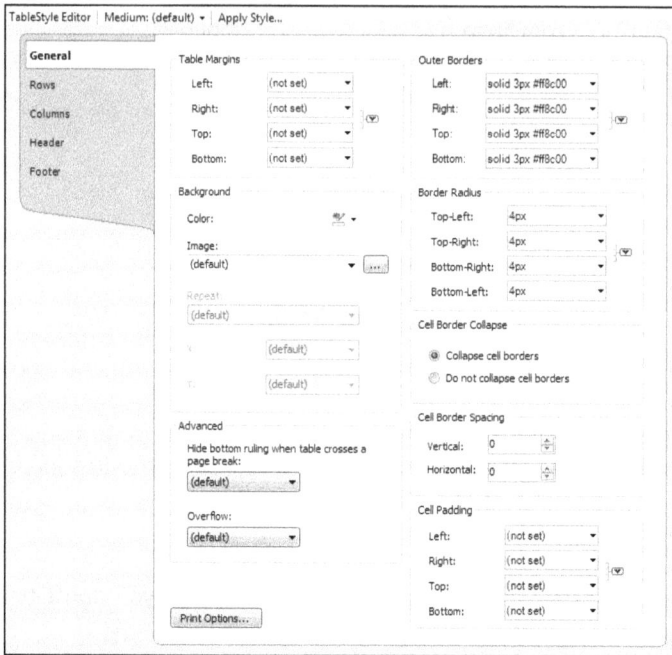

TableStyle Editor | Medium: (default) ▾ | Apply Style...

General	Table Margins			Outer Borders		
Rows	Left:	(not set) ▾		Left:	solid 3px #ff8c00 ▾	
Columns	Right:	(not set) ▾		Right:	solid 3px #ff8c00 ▾	
Header	Top:	(not set) ▾		Top:	solid 3px #ff8c00 ▾	
Footer	Bottom:	(not set) ▾		Bottom:	solid 3px #ff8c00 ▾	

Background
Color: ▾
Image:
(default)   ▾ [...]
Repeat
(default)   ▾
(default)   ▾
(default)   ▾

Border Radius
Top-Left: 4px   ▾
Top-Right: 4px   ▾
Bottom-Right: 4px   ▾
Bottom-Left: 4px   ▾

Cell Border Collapse
◉ Collapse cell borders
○ Do not collapse cell borders

Cell Border Spacing
Vertical: 0
Horizontal: 0

Advanced
Hide bottom ruling when table crosses a page break:
(default)   ▾

Overflow:
(default)   ▾

Cell Padding
Left: (not set) ▾
Right: (not set) ▾
Top: (not set) ▾
Bottom: (not set) ▾

[Print Options...]

2   On the **General** tab, select the following options:

□ **Background** — select a background color or image.

□ **Border Radius** — select the radius for curved table border corners (or select 0 for squared corners).

□ **Cell Border Collapse** — cell borders normally appear inside row borders. If you collapse them, they are merged with the row border.

□ **Cell Border Spacing** — select the amount of space between cells.

□ **Cell Padding** — select the amount of space between a cell's border and its content.

□ **Outer Borders** — select the style, width, and color of your table borders.

□ **Table Margins** — select the amount of space between the table and the content around the table.

3 Select the **Rows** tab.

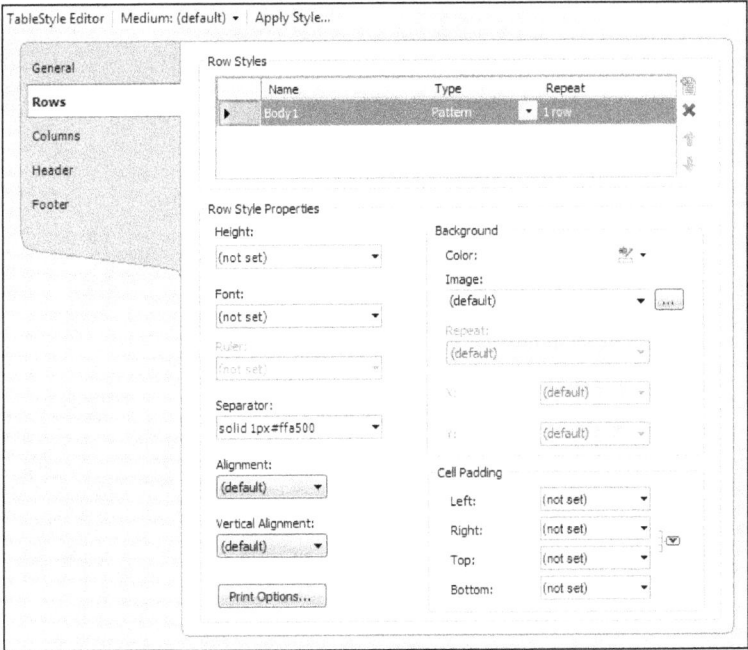

TableStyle Editor | Medium: (default) ▾ | Apply Style...

Row Styles

	Name	Type	Repeat	
▶	Body 1	Pattern ▾	1 row	

Row Style Properties

Height:
(not set) ▾

Font:
(not set) ▾

Ruler:
(not set) ▾

Separator:
solid 1px #ffa500 ▾

Alignment:
(default) ▾

Vertical Alignment:
(default) ▾

Print Options...

Background

Color: ▾

Image:
(default) ▾ [...]

Repeat:
(default) ▾

X: (default) ▾

Y: (default) ▾

Cell Padding

Left: (not set) ▾

Right: (not set) ▾

Top: (not set) ▾

Bottom: (not set) ▾

4 On the **Rows** tab, select the following options:

- **Row Styles** — patterns can be used to provide different row formats, such as alternating background colors.

- **Row Style Properties** — if you use a pattern, select how many times the pattern should repeat, its background color, text color, and a separator border.

5 Select the **Columns** tab.

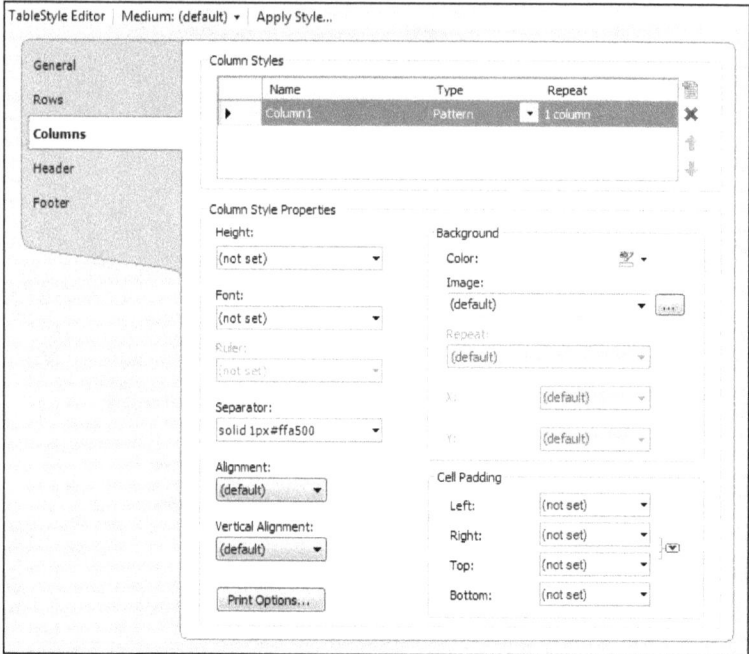

6 On the **Columns** tab, select the following options:

- ☐ **Column Styles** — patterns can be used to provide different column formats, such as alternating background colors.

- ☐ **Column Style Properties** — if you use a pattern, select how many times the pattern should repeat, its background color, text color, and a separator border.

7 Select the **Header** tab.

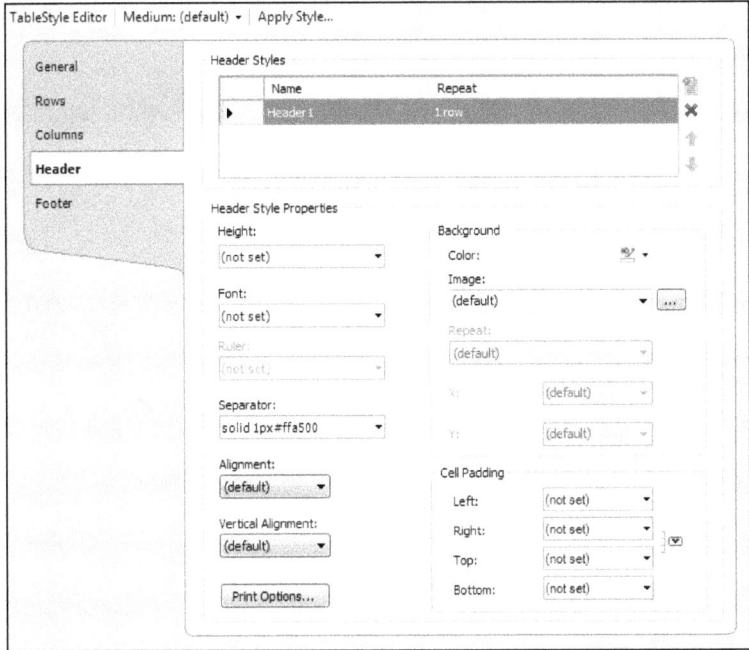

TableStyle Editor | Medium: (default) ▾ | Apply Style...

General	**Header Styles**	

	Name	Repeat	
▸	Header 1	1 row	✖

Rows

Columns

Header

Footer

Header Style Properties

Height:
(not set) ▾

Font:
(not set) ▾

Ruler:
(not set) ▾

Separator:
solid 1px#ffa500 ▾

Alignment:
(default) ▾

Vertical Alignment:
(default) ▾

Print Options...

Background

Color: ⬚ ▾

Image:
(default) ▾ ...

Repeat:
(default) ▾

X: (default) ▾

Y: (default) ▾

Cell Padding

Left: (not set) ▾

Right: (not set) ▾

Top: (not set) ▾

Bottom: (not set) ▾

8 On the **Header** tab, select the following options:

☐ **Header Styles** — patterns can be used to provide different header formats, such as a bottom border.

☐ **Header Style Properties** — if you use a pattern, select how many times the pattern should repeat, its background color, text color, and a separator border.

9 Select the **Footer** tab.

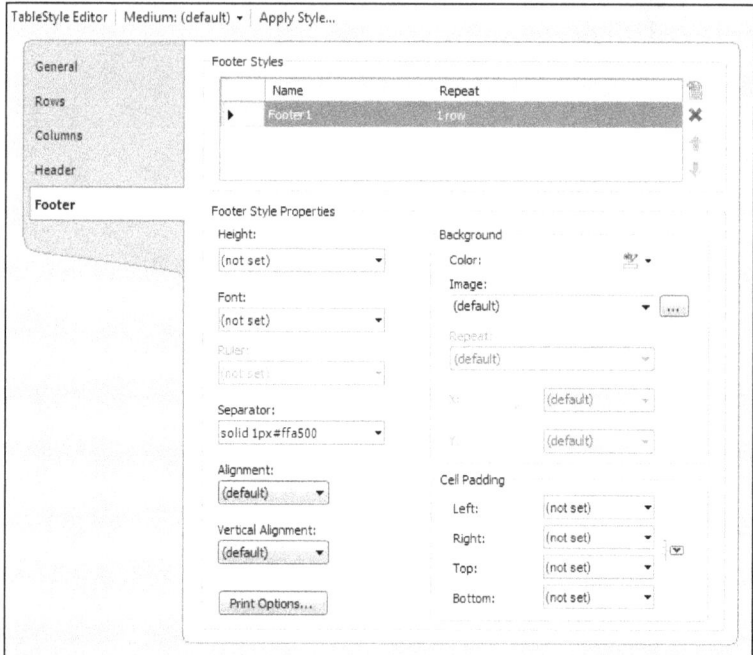

TableStyle Editor | Medium: (default) ▾ | Apply Style...

General
Rows
Columns
Header
Footer

Footer Styles

	Name	Repeat
▶	Footer 1	1 row

Footer Style Properties

Height:
(not set)

Font:
(not set)

Ruler:
(not set)

Separator:
solid 1px#ffa500

Alignment:
(default)

Vertical Alignment:
(default)

Print Options...

Background
Color:

Image:
(default)

Repeat:
(default)

x:
(default)

y:
(default)

Cell Padding
Left: (not set)
Right: (not set)
Top: (not set)
Bottom: (not set)

10 On the **Footer** tab, select the following options:

☐ **Footer Styles** — patterns can be used to provide different footer formats, such as a top border.

☐ **Footer Style Properties** — if you use a pattern, select how many times the pattern should repeat, its background color, text color, and a separator border.

Specifying a cell content style

Table cells use the th (table header), td (table data), or tf (table footer) tags. If you add a paragraph to a cell, the p (paragraph) tag is added to the cell. If you have formatted the p tag in your stylesheet, the table cells that contain p tags could have different formatting than the cells without p tags.

To ensure consistent formatting for cells with or without p tags, you can specify a cell content style. For example, you can set Flare to automatically add a p tag to every cell when you apply a table style.

To specify a cell content style:

1 If you want to use a class for your header, body, or footer cell content, create a class in your stylesheet:

   ☐ Open your stylesheet.

   ☐ Select the **p** tag.

   ☐ Click **Add Class**.

   ☐ Type a name for the class.

   ☐ Click **OK**.

   ☐ Specify the formatting you want to use for the cell content.

2 Open a table style.

3 Select the **Header** tab.

4 In the **Cell Content Style** group box, select the **p** tag.

5 If you create a class for header cells, select the class.

6 Select the **Row** or **Column** tab.

7 In the **Cell Content Style** group box, select the **p** tag.

8 If you create a class for header cells, select the class.

9 Select the **Footer** tab.

10 In the **Cell Content Style** group box, select the **p** tag.

11 If you create a class for header cells, select the class.

Applying a table style to a table

After you create a table style, you can apply it to specific tables or all topics in a file, folder, or project.

Shortcut	Tool Strip	Ribbon
Alt+B, P	Table > Table Properties	Table > Apply Table Style

To apply a table style to a table:

1   Click inside the table.

2   Select **Table > Table Properties.**
The Table Properties dialog box appears.

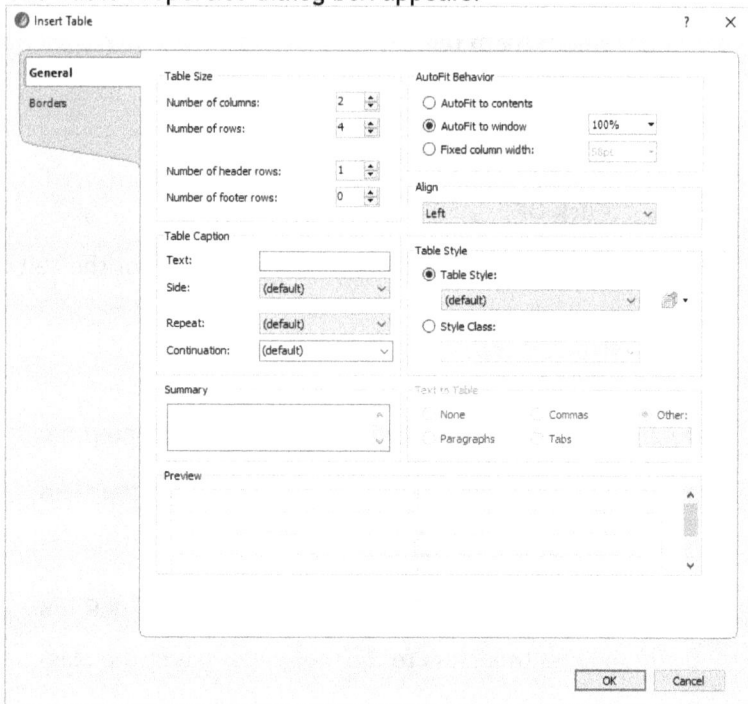

3   Select a **Table Style.**

4   Click **OK.**

To apply a table style to multiple tables:

1   Open a table stylesheet.

2   Click **Apply Style.**
The Apply Table Style dialog box appears.

3 Select a topic or folder. If you select a topic, the table style will be applied to all of the tables in the selected topic. If you select a folder, the table style will be applied to all of the tables in the topics stored in the folder.

> **TIP** *To apply the table style to all of the tables in your project, select the Content folder.*

4 Select **Overwrite existing table styles** if you want to apply the table style to tables that currently use another table style.

5 Select **Remove local formatting** if you want to also remove inline formatting from your tables.

6 Click **OK**.

Removing inline formatting from a table

If a table contains inline formatting, the inline formatting will override the formatting in your table style. For example, if your table contains inline formatting that adds red borders, the borders will be red even if the table stylesheet specifies black borders.

Inline table formatting is often found in topics that are imported from RoboHelp or Word.

Shortcut	Tool Strip and Ribbon
Alt+B, S, F	Table > Reset Local Cell Formatting

To remove inline formatting from a table:

1 Select **Table** > **Select Table.**

2 Select **Table** > **Reset Local Cell Formatting.**

Deleting a table, row, column, or content

You can press **Backspace** to delete an entire table, table columns, or table rows. If you want to delete content inside a table, you can press **Delete.**

To delete a table:

1 Select the table.

2 Press **Backspace.**

To delete a table row or column:

1 Select the rows/columns.

2 Press **Backspace.**

To delete content in a table:

1 Select the rows/columns.

2 Press **Delete.**

Images and multimedia

You can use any of the following image, video, and sound file types:

Image file types

- bmp
- emf (or wmf)
- **eps** (or ps)
- **gif**
- hdp (or wdp)
- **jpg** (or jpeg)
- **png**
- **svg**
- tif (or tiff)
- wdp
- xaml
- xps (or exps)

Video file types

- asf (or asx)
- avi
- mov
- **mp4**
- mpg (or mpeg)
- mimov, miprj, mcmovie, mcmoviesys, and mcmv
- ogg (or ogv)
- qt
- swf
- u3d
- **webm**
- wmv

Sound file types

- au
- midi (or mid)
- **mp3**
- opus
- wav
- wma

The most popular formats are in bold.

TIP *Capture, MadCap's image editing application, integrates very well with Flare.*

Inserting an image

You can insert images into topics, snippets, template pages, and page layouts. By default, images are stored in the Resources\Images folder.

TIP *You can create a style to display images as small "thumbnails" that users can enlarge when needed. See "Creating an image thumbnail style" on page 238.*

Shortcut	Tool Strip	Ribbon
Ctrl+G	🖼 (XML Editor)	Insert > Image

To insert an image:

1 Open a file.

2 In the XML Editor, position your cursor where you want to insert the image.

3 Click 🖼 in the XML Editor toolbar.
   —OR—
   Select **Insert** > **Image**.
   The Insert Image dialog box appears.

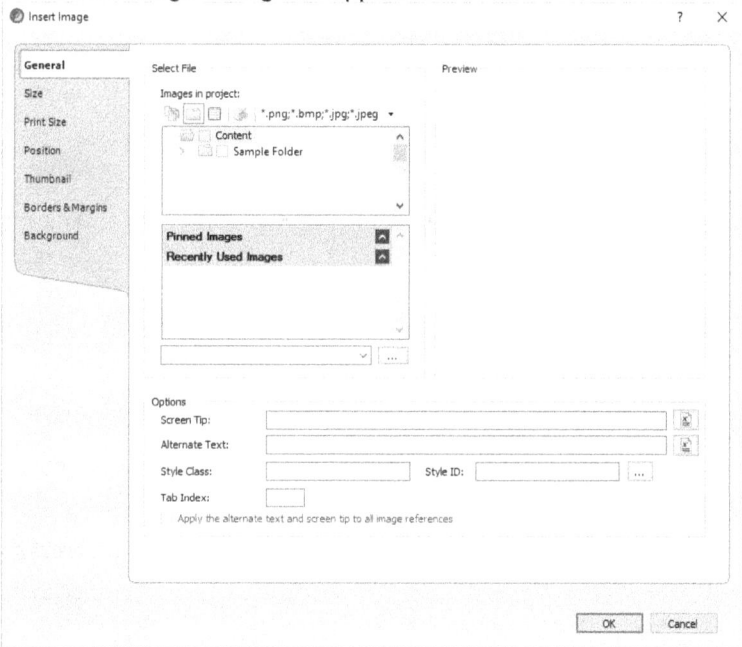

4 Select an image in the Images in Project list.
   —OR—
   Click 🔲 and select an image.

5 Type an **Alternate Text** description of the link.
Alt text is recommended by accessibility guidelines such as the US Government's Section 508 and the W3C's Web Content Accessibility Guidelines (WCAG).

6 Click **OK**.

Inserting a PDF page as an image

You can insert a page from a PDF as an image. If you build a PDF target, the PDF image becomes a regular page in the PDF. In other targets, the PDF image becomes a PNG image.

TIP> *You can create a style to display images as small "thumbnails" that users can enlarge when needed. See "Creating an image thumbnail style" on page 238.*

Shortcut	Tool Strip	Ribbon
Ctrl+G	▦ (XML Editor)	Insert > Image

To insert an image:

1 Open a file.

2 In the XML Editor, position your cursor where you want to insert the PDF image.

3 Click ▦ in the XML Editor toolbar.
—OR—
Select **Insert** > **Image**.
The Insert Image dialog box appears.

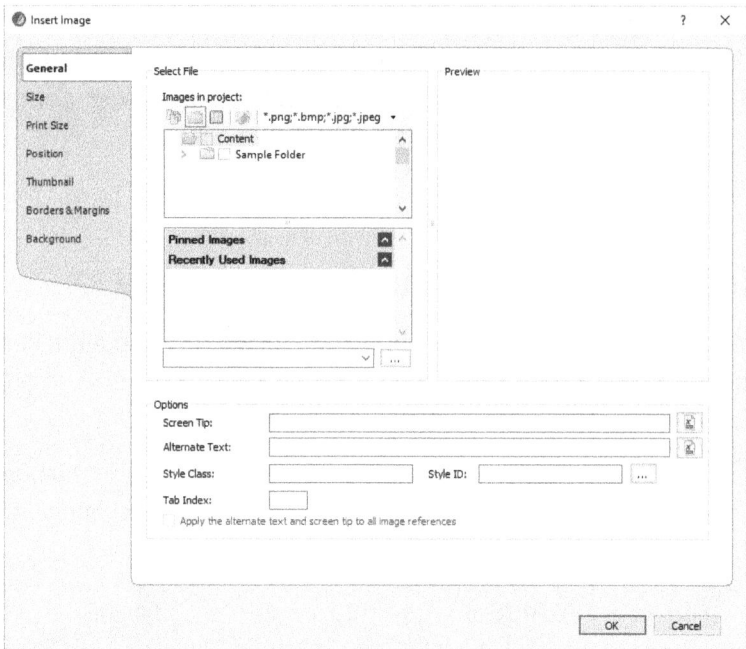

4 Click [...] and select a PDF document.
The PDF appears in the Preview box.

5 Select a **Page**.

6 Type an **Alternate Text** description of the link.
Alt text is recommended by accessibility guidelines such as the
US Government's Section 508 and the W3C's Web Content
Accessibility Guidelines (WCAG).

7 Click **OK**.

Inserting a video

You can insert Flash (swf), Windows Media Player (asf and mpg),
QuickTime (mov, mp4, and qt), Mimic, and HTML5 (webm, ogg, and
mp4) videos into topics, snippets, and template pages.

By default, videos are stored in the Resources\Multimedia folder.

Shortcut	Tool Strip and Ribbon
Alt+N, M	Insert > Multimedia

To insert a video:

1 Open a file.

2 In the XML Editor, position your cursor where you want to insert the movie.

3 Select **Insert** > **Multimedia** and select either:

- ▫ **Flash** — swf movies

- ▫ **Windows Media Player** — asf or mpg movies

- ▫ **QuickTime** — mov or qt movies

- ▫ **Mimic** — MadCap Mimic mimov, miprj, mcmovie, mcmoviesys, and mcmv movies

- ▫ **HTML5** — mp4, ogg, or webm movies

The Insert Multimedia dialog box appears.

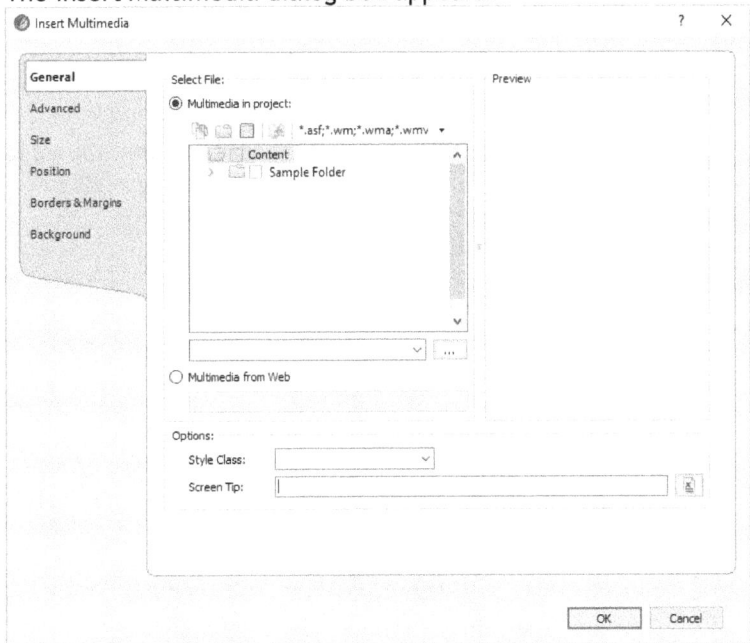

4 Select a movie in the Multimedia in Project list.
—OR—
Click ⌐··⌐ and select a movie.

5 Type a screen tip that describes the movie.
Screen tips are recommended by accessibility guidelines such as the US Government's Section 508 and the W3C's Web Content Accessibility Guidelines (WCAG).

6 Select the **Advanced** tab.

7 Select whether the video loops, auto starts, and/or provides playback controls.

◇ *The advanced options vary based on the video type.*

8 Click **OK**.
The movie appears in your file as a grey box, but it will play normally in the preview and in your targets.

Inserting a YouTube or Vimeo video

You can insert YouTube or Vimeo videos into topics, snippets, and template pages.

TIP▷ *You can also create a link to a YouTube or Vimeo video. See "Creating a hyperlink to a website, external file, or email address" on page 154.*

Shortcut	Tool Strip and Ribbon
Alt+N, M	Insert > Multimedia

To insert a YouTube or Vimeo video:

1 Open a file.

2 In the XML Editor, position your cursor where you want to insert the movie.

3 Select **Insert** > **Multimedia** > **YouTube/Vimeo**.
The Insert Multimedia dialog box appears.

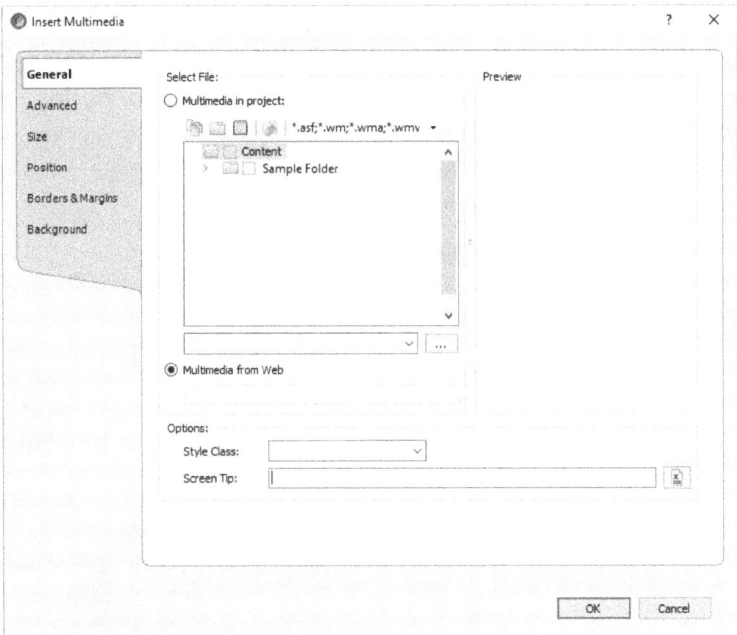

4 Select **Multimedia from Web**.

5 Type or paste the video's URL.

6 Type a screen tip that describes the movie.
Screen tips are recommended by accessibility guidelines such as the US Government's Section 508 and the W3C's Web Content Accessibility Guidelines (WCAG).

7 Select the **Advanced** tab.

8 Set the playback options, including whether the video auto plays and/or loops.

9 Click **OK**.
The movie appears in your file as a grey box, but it will play normally in the preview and in your targets.

Inserting a 3D model

You can insert universal 3D (U3D) models to provide interactive three-dimensional images.

◇ *3D models (U3D files) are not supported in EPUB or MOBI files, and they require a plug-in and other settings to display correctly in browsers and PDF files. For the most up-to-date information, see "Inserting 3D Models" in the Flare help system.*

Shortcut	Tool Strip and Ribbon
Alt+N, M	Insert > Multimedia

To insert a 3D model:

1 Open a file.

2 In the XML Editor, position your cursor where you want to insert the 3D model.

3 Select **Insert > Multimedia > 3D Model**.
   The Insert Multimedia dialog box appears.

4 Select a movie in the Multimedia in Project list.
   —OR—
   Click 🔲 and select a movie.

5 Type a screen tip that describes the 3D model.
Screen tips are recommended by accessibility guidelines such as the US Government's Section 508 and the W3C's Web Content Accessibility Guidelines (WCAG).

6 Select the **Advanced** tab.

7 Set the **Activation** options, including when the 3D model activates.

8 Set the **Rendering** options, including the model's lighting and background color.

9 Click **OK**.
The 3D model appears in your file as a grey box, but it will play normally in the preview and in your targets.

Inserting a sound

You can insert sounds into topics, snippets, and template pages. By default, sounds are stored in the Resources\Multimedia folder.

Shortcut	Tool Strip and Ribbon
Alt+N, M	Insert > Multimedia

To insert a sound:

1 Open a file.

2 In the XML Editor, position your cursor where you want to insert the sound.

3 Select **Insert >Multimedia > Windows Media Player**.
The Insert Multimedia dialog box appears.

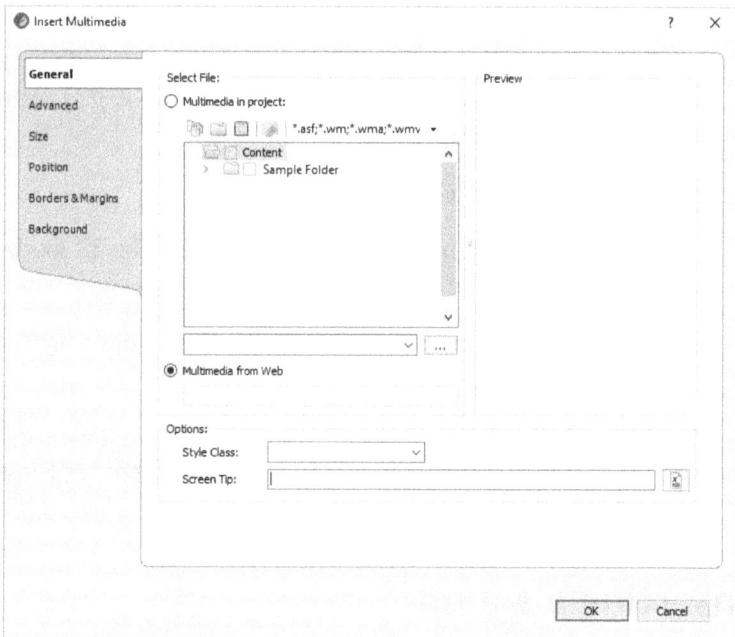

4 Select a sound in the Multimedia in Project list.

—OR—

Click ⌑ and select a sound.

5 Click **OK**.

The sound appears in your file as a grey box, but it will play normally in the preview and in your targets.

Pinning an image

You can pin frequently used images in the Insert Image dialog box. Pinning an image makes it easier to find.

To pin an image in the images list:

1 Select **Insert > Image**.

The Insert Image dialog box appears.

2 In the **Recently Used Images** box, hover over the image.

3 Click ▪.

The image is pinned.

'How do I see which files use an image, video, or sound?'

You can right-click any multimedia file and select **View Links** to view a list of topics that include it. In fact, you can right-click any file, including stylesheets, topics, and PDF documents, to see a list of files that include or link to it.

Slideshows

You can insert a slideshow to allow users to browse content in a specified order. For example, a slideshow could be used to explain stages in a process or steps in a workflow. Or, you could create a slideshow to provide a graphical list of links. The Flare help system uses a slideshow to provide links to topic-specific PDF guides.

◇ *Slideshows can only be used in online and EPUB targets. If you are also creating another type of print target, you should apply a condition tag to the slideshow and exclude it from your print target. See "Applying a tag to content" on page 315. If you are creating an EPUB target, you can enable or disable dynamic content such as slideshows. See 'Setting up a print target' on page 338.*

Inserting a slideshow

You can insert a slideshow into a topic or snippet.

◇ *You cannot insert a slideshow inside a drop-down or toggler.*

Shortcut	Tool Strip and Ribbon
Alt+N, S, S	Insert > Slideshow

To insert a slideshow:

1 Open a file.

2 In the XML Editor, position your cursor where you want to insert the slideshow.

3 Select **Insert > Slideshow**.

Adding a slide to a slideshow

When you create a slideshow, it contains two sample slides. You can modify or delete the sample slides and add additional slides. A slide can contain any content, including text, images, and videos.

To add a slide to a slideshow:

1   Click ✛.
    —OR—
    Right-click the slideshow tag in the block bar and select **Add Slide**.

2   Click inside the new slide and add your content.

To delete a slide from a slideshow:

1   Click ← or → to select the slide.

2   Click ✖.

Formatting a slideshow

You can set the slideshow's navigation options and each slides' caption, thumbnail icon, and position in the slideshow.

To format a slideshow:

1   Right-click the slideshow and select **Edit Slideshow**.
    The Edit Slideshow dialog box appears.

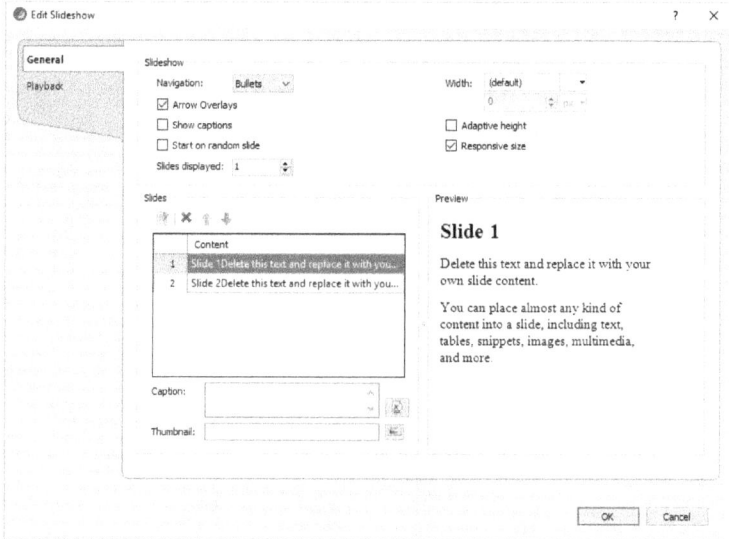

2 Select the **General** tab.

3 Select a **Navigation** option.

4 If you want to provide navigational arrows, select **Arrow Overlays**.

If you don't provide arrows, users can select slides by clicking the thumbnail icons below the slides.

5 If you want to include slide captions:

- □ Select **Show Captions**.

- □ Select each slide in the **Slides** group box.

- □ Type a caption.

6 If you want to display more than one slide at a time, set a **Slides displayed** number.

7 If you want to include slide thumbnails:

- □ Select a slide.

- □ Click .

- □ Select a thumbnail icon and click **OK**.

8 Select the **Playback** tab.

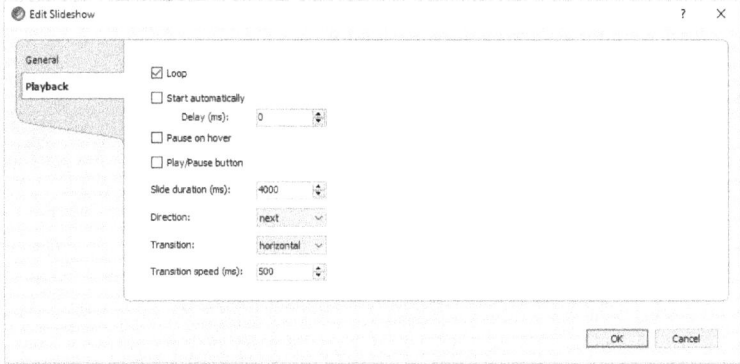

9 Set the playback options.
You can specify whether the slideshow loops, starts automatically, pauses when the user hovers over a slide, and other options.

10 Click **OK**.

Sample questions for this section

1 Which of the following statements is true?
   A) You can edit HTML and XHTML topics in Flare.
   B) You must know HTML or XML to use Flare.
   C) XHTML is an XML schema.
   D) All of the above.

2 How many topics can you have open in Flare?
   A) One
   B) Two (one in the XML editor and one in the Internal Text Editor)
   C) Up to twenty
   D) As many as you want.

3 What does the chain (⬚) icon indicate when it appears after the filename when you open a topic?
   A) The topic is linked to an HTML, Word, or FrameMaker document.
   B) The topic needs to be saved.
   C) The topic is linked to a template.
   D) The topic is connected to source control.

4 You can import and link the following types of Flare files between projects:
   A) Topics
   B) Stylesheets
   C) Page layouts
   D) All of the above

5 How do you import a PDF file?
   A) Right-click the **Imports** folder and create a PDF import file.
   B) Save the PDF as HTML and import the HTML file.
   C) Select **File > Import > PDF File.**
   D) Select **Project > Add PDF File.**

6 Table style files have the following extension:
   A) .htm
   B) .tss
   C) .css
   D) .fltbl

7   Which of the following image types can you NOT insert into a topic?

A) jpg

B) svg

C) eps

D) ai

8   How can you view a list of topics that include an image?

A) Select **View** > **Image List**.

B) Right-click the image in the Content Explorer and select **View Links**.

C) Hover your cursor over the image in a topic.

D) Open the Images Analyzer report in the **Analysis** menu.

Links

This section covers:

- Hyperlinks
- Popup links
- Cross references
- Drop-down, expanding, and toggler links
- Related topic, keyword, and concept links
- Relationship links

Hyperlinks

You can create hyperlinks that open:

- ☐ Topics
- ☐ Bookmarks in topics
- ☐ Documents such as .doc, .xls, and .ppt files
- ☐ PDF files (or even destinations inside a PDF)
- ☐ Websites
- ☐ Email messages

You can add hyperlinks anywhere in a topic. They are often included at the end of a topic to suggest related topics. A link's text should clearly identify what will happen when it is clicked. For example:

Email technical support (well worded link label)

Click here if you have a question (poorly worded link label)

Unvisited links are usually blue and underlined, and visited links are usually purple and underlined. You can change their appearance by modifying the "a" (for "anchor") style in your stylesheet.

TIP *If you Ctrl-click a link in the XML Editor, the linked topic, document, or website will open as a new tab in the XML Editor.*

Creating a hyperlink to a topic

Shortcut	Tool Strip	Ribbon
Ctrl+K	(XML Editor)	Insert > Hyperlink

To create a hyperlink to a topic:

1 Open the topic that will contain the link.

2 Highlight the text that you want to use as the link.
—OR—
Right-click an image and select **Select**.

3 Select **Insert** > **Hyperlink**.
—OR—
Right-click and select **Hyperlink**.
The Insert Hyperlink dialog box appears.

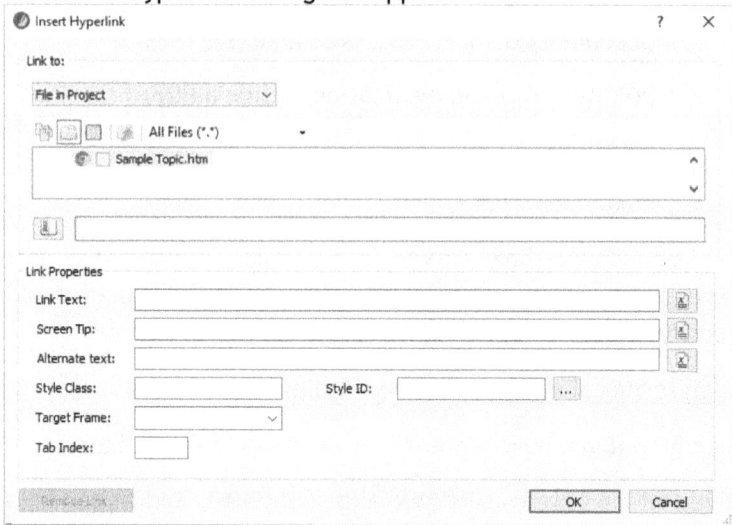

4 In the **Link to** section, select **File in Project**.

5 Select a topic.
TIP *You can click the ⬚ icon to view the topics organized by folder rather than file name.*

6 If you want to link to a location in a topic, such as a bookmark, heading, list item, or a style:

- Click ⬚.

- Select a location.

- Click **OK**.

7 Type the **Alternate Text**.
Alternate text is recommended by accessibility guidelines such as Section 508 of the U.S Government's Rehabilitation Act and the W3C's Web Content Accessibility Guidelines (WCAG).

8 If needed, select a **Style Class** for the link.

9 Select a **Target Frame**.
The target frame specifies where the link will appear. For example, you can select "New Window" to open the link in a new window. By default, the link will open in the current window.

10 Click **OK**.
The hyperlink is added to the topic.

Creating a hyperlink to a PDF

Shortcut	Tool Strip	Ribbon
Ctrl+K	(XML Editor)	Insert > Hyperlink

To create a hyperlink to a PDF document:

1 Locate the document to which you want to link.

2 Drag-and-drop or copy the document into the Content folder.
If you want to leave the document in its current location, see "Creating a hyperlink to a website, external file" on page 154.

3 Open the topic that will contain the link.

4 Highlight the text that you want to use as the link.
—OR—
Click an image and select **Select**.

5 Click in the XML Editor toolbar.
—OR—
Right-click and select **Hyperlink**.
The Insert Hyperlink dialog box appears.

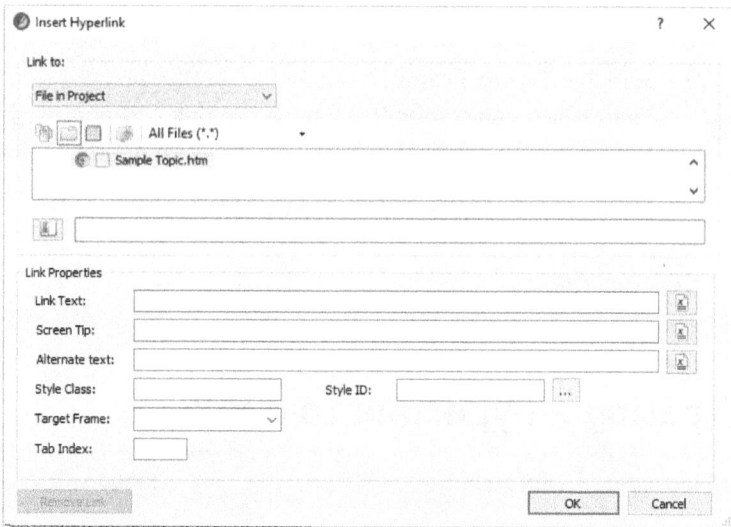

6 In the **Link to** section, select **File in Project**.

7 Select a document. If you do not see your document in the list, you may need to change the filter setting to "All Files."

8 If you want to link to a destination in the PDF:

☐ Click .

☐ Select a destination.

☐ Click **OK**.

9 Type the **Alternate Text**.
Alternate text is recommended by accessibility guidelines such as Section 508 of the U.S Government's Rehabilitation Act and the W3C's Web Content Accessibility Guidelines (WCAG).

10 If needed, select a **Style Class** for the link.

11 Select a **Target Frame**.
The target frame specifies where the link will appear. For example, you can select "New Window" to open the link in a new window. By default, the link will open in the current window.

12 Click **OK**.
The hyperlink is added to the topic.

Creating a hyperlink to a DOC, PPT, or XLS file

Shortcut	Tool Strip	Ribbon
Ctrl+K	▥ (XML Editor)	Insert > Hyperlink

To create a link to a Word, PowerPoint, or Excel document:

1 Locate the document to which you want to link.

2 Drag-and-drop or copy the document into the Content folder. If you want to leave the document in its current location, see "Creating a hyperlink to a website, external file" on page 154.

   ◇ *If you drag-and-drop a Word document into the Content Explorer, Flare will assume you want to import the Word document. However, you can drag-and-drop it into the Content folder in Windows.*

3 Open the topic that will contain the link.

4 Highlight the text that you want to use as the link.
   —OR—
   Click an image and select **Select**.

5 Click ▥ in the XML Editor toolbar.
   —OR—
   Right-click and select **Hyperlink**.
   The Insert Hyperlink dialog box appears.

6 In the **Link to** section, select **File in Project**.

7 Select a document. If you do not see your document in the list, you may need to change the filter setting to "All Files."

8 Type the **Alternate Text**.
Alternate text is recommended by accessibility guidelines such as Section 508 of the U.S Government's Rehabilitation Act and the W3C's Web Content Accessibility Guidelines (WCAG).

9 If needed, select a **Style Class** for the link.

10 Select a **Target Frame**.
The target frame specifies where the link will appear. For example, you can select "New Window" to open the link in a new window.

11 Click **OK**.
The hyperlink is added to the topic.

Creating a hyperlink to a website, external file, or email address

Shortcut	Tool Strip	Ribbon
Ctrl+K	(XML Editor)	Insert > Hyperlink

To create a link to a website or external file:

1 Open the topic that will contain the link.

2 Highlight the text that you want to use as the link.
—OR—
Click an image and select **Select**.

3 Click 🖳 in the XML Editor toolbar.
—OR—
Right-click and select **Hyperlink**. The Insert Hyperlink dialog box appears.

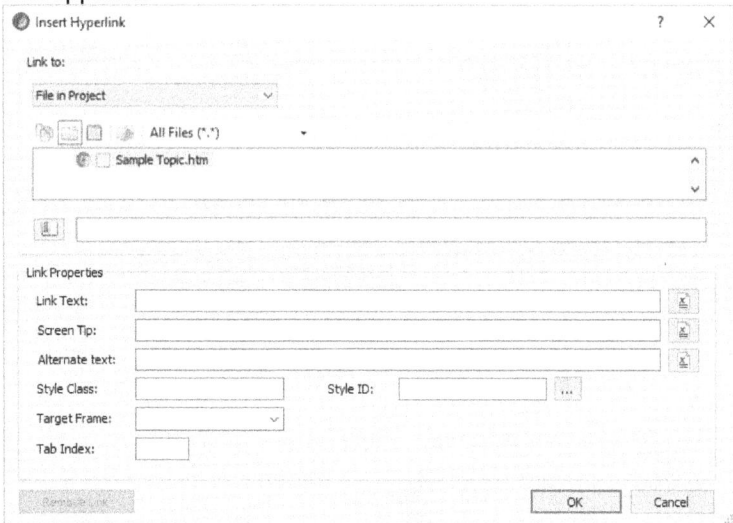

4 In the **Link to** section, select **Website**.

5 Type the path to the website or document. To create an email link, type "mailto:" before the email address.

6 Type the **Alternate Text**.
Alternate text is recommended by Section 508 of the U.S Government's Rehabilitation Act and the W3C's Web Content Accessibility Guidelines (WCAG).

7 If needed, select a **Style Class** for the link.

8 Select a **Target Frame**.
Links to websites often appear in a new window.

9 Click **OK**.

The hyperlink is added to the topic.

Creating an image map link

An image map allows you to add links within an image. You can add as many links as needed to an image, and the links can be rectangular, oval, or irregular shapes. For example, you could add links to a picture of the United States so that each state was a link.

You can include image maps and other dynamic content such as drop-downs in PDF targets. For more information, see "Setting up a print target" on page 338.

To create an image map link:

1 Open the topic that contains the picture to which you want to add links.

2 Right-click the image and select **Image Map**.
The Image Map Editor window appears.

3 If the image appears faded, click .

4 Select an image map shape and draw your shape.

5 Double-click your shape.
The Area Properties dialog box appears.

6 In the **Link to** section, select a link target type, such as a topic or website, and a target.

7 Type the **Alternate Text**.
Alternate text is recommended by accessibility guidelines such as Section 508 of the U.S Government's Rehabilitation Act and the W3C's Web Content Accessibility Guidelines (WCAG).

8 Select a **Target Frame**.

9 Click **OK**.

10 Click **OK** in the Image Map Editor toolbar.
The Image Map Editor closes, and the image map is added to your image.

To apply a condition tag to an image map link:

1 Right-click the image map link and select **Conditions**.

2 Select a condition tag.

3 Click **OK**.

'How do I view a topic's links?'

You can view a list of links by opening a topic (or any type of file) and selecting **View** > **Link Viewer**.

Finding and fixing broken links

You can use the Broken Links report to find and fix broken links in your project. Broken links are not a common problem, but they can occur when you import content with broken links or if you delete a topic and don't click **Remove Links**.

To find and fix broken links:

1 Select **Analysis** > **Links** > **Broken Links**.
   A list of broken links appears.

2 Double-click a broken link in the list.
   The topic opens, and the broken link is highlighted.

3 Right-click the highlighted link and select **Edit Hyperlink**.

4 Select a new link location.

5 Click **OK**.

Finding and testing links to websites

You can use the External Links report to find and test hyperlinks to websites.

To find and test website hyperlinks:

1 Select **Analysis** > **Links** > **External Links**.
   The External Links report appears.

2 Highlight the hyperlink(s) you want to test.

3 Click **Check**.

Popup links

You can create two types of popup links: topic popups and text popups.

Topic popups are links that open another topic in a popup window. Since a popup link opens another topic, the popup content can contain formatted text, images, tables, and lists.

Text popups are links that display hidden text in a popup window. They can only contain unformatted text. Since the popup's content is hidden inside the topic that contains the link, you cannot reuse a text popup in multiple topics. Instead, you must retype the content in each topic.

Creating a topic popup link

You can create topic popup links to open topics or other documents in a popup window. They are often used to provide definitions for terms and acronyms.

Like a "normal" link, a topic popup link opens another topic. The difference is that the topic opens in a popup window that closes when it loses focus. A normal link can open in a new window, but the new window will not automatically close. Flare can automatically size the popup window based on the popup's content, or you can specify the width and height in your stylesheet.

Popup links are also similar to drop-down, expanding, and toggler links. However, drop-down, expanding, and toggler links show and hide content in the current topic rather than in a popup window. Popup links are not as popular as these other link types because they can cause problems with popup blockers and because they can be hard to print. See "Drop down, expanding, and toggler links" on page 165.

Shortcut	Tool Strip and Ribbon
Alt+N, I, P	Insert > Topic Popup

To create a topic popup link:

1 Open the topic that will contain the link.

2 Highlight the text that you want to use as the link.
   —OR—
   Click an image and select **Select**.

3 Select **Insert > Hyperlink > Topic Popup**.
   The Insert Topic Popup dialog box appears.

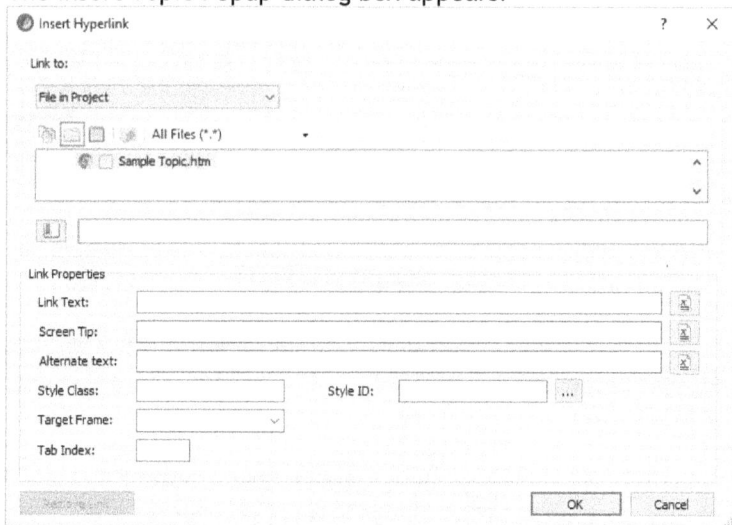

4 In the **Link to** section, select a link target type, such as a topic or website, and select a link target.

5 Type the **Alternate Text**.
   Alternate text is recommended by accessibility guidelines such as Section 508 of the U.S Government's Rehabilitation Act and the W3C's Web Content Accessibility Guidelines (WCAG).

6 Click **OK**.
   The popup link is added to the topic.

Creating a text popup link

Text popups can only display unformatted text. The popup's content is stored inside the topic that contains the link, so you cannot reuse a text-only popup in another topic without retyping the popup's content.

✏ Text popups can only be used in online and EPUB targets. If you are creating an EPUB target, you can enable or disable dynamic content such as text popups. See 'Setting up a print target' on page 338.

Shortcut	Tool Strip and Ribbon
Alt+N, P, U	Insert > Text Popup

To create a text popup link:

1 Open the topic that will contain the link.

2 Highlight the text that you want to use as the link.
   —OR—
   Click an image and select **Select**.

3 Select **Insert** > **Text Popup**.
   The Insert Text Popup dialog box appears.

4 Type the popup text.

5 Click **OK**.
   The popup link appears.

Cross references

Cross reference links provide two advantages over "normal" hyperlinks:

- The link's label can use a variable to include the target topic's title

- The link's label can include page numbers for print targets

If a cross reference's link label uses a variable to include the target topic's title, the label will be updated automatically if you change the topic's title.

Cross reference labels can be set up using a cross reference style named "MadCap|xref." You can set up the style to automatically add words, format, or add page numbers to your cross reference labels.

Creating a cross reference

Shortcut	Tool Strip	Ribbon
Ctrl+Shift+R	(XML Editor)	Insert > Cross Reference

To create a cross reference:

1 Open a topic.

2 Position your cursor where you want to insert the cross reference.

3 Click in the XML Editor toolbar.
   —OR—
   Right-click and select **Cross Reference**.

The Insert Cross Reference dialog box appears.

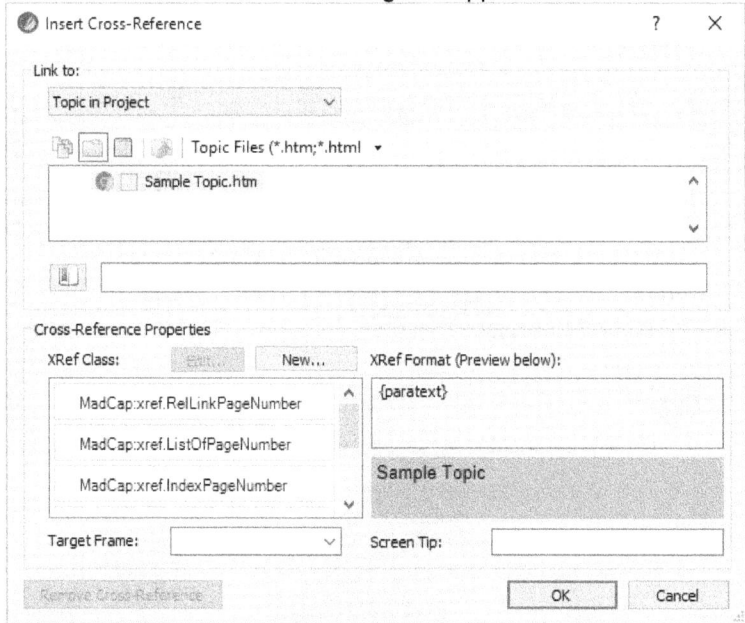

4 For **Link To**, select **Topic in Project**.

5 Select a topic.
   TIP *You can click the* 🗁 *icon to view the topics organized by folder rather than file name.*

6 If you want to link to a location in a topic, such as a bookmark, heading, list item, or a style:

   ☐ Click 🔲.

   ☐ Select a location.

   ☐ Click **OK**.

7 Type the **Alternate Text**.
   Alternate text is recommended by accessibility guidelines such as Section 508 of the U.S Government's Rehabilitation Act and the W3C's Web Content Accessibility Guidelines (WCAG).

8 Click **OK**.

To create a cross reference using drag-and-drop:

1 Open a topic.

2 Select a topic in the Content Explorer, drag it into the XML Editor, and release the mouse button.

Converting hyperlinks to cross references

You can convert hyperlinks to cross references to take advantage of their advanced link label features.

To convert hyperlinks to cross references:

1 Select **Analysis** > **Suggestions** > **Cross Reference Suggestions**. The Cross Reference Suggestions report appears.

2 Highlight the hyperlink(s) you want to convert.

3 Click .

Drop-down, expanding, and toggler links

You can create three types of "show/hide" links: drop-down, expanding, and toggler. By default, these links can include expanded/collapsed arrow icons. You can change or remove these icons by modifying the link styles.

◇ *Drop-down, expanding, and toggle links can only be used in online and EPUB targets. If you are creating an EPUB target, you can enable or disable these layering links. See 'Setting up a print target' on page 338.*

'What's the difference?'

A **drop-down** link shows and hides a paragraph, image, or list item *below* the drop-down link. Drop-down links are often used to show and hide content between subheadings.

An **expanding** link shows and hides a word or sentence *within* a paragraph or list item. Expanding links are often used to show and hide short definitions.

A **toggler** link shows and hides a named element (such as a paragraph, image, or list item) *anywhere* in a topic. Toggler links are often used to show and hide a screenshot or table from a link at the top of a topic.

Creating a drop-down link

Shortcut	Tool Strip and Ribbon
Alt+N, D	Insert > Drop-Down Text

To create a drop-down link:

1 Open the topic that will contain the drop-down link.

2 Type and highlight the drop-down link and drop-down text.

3 Select **Insert > Drop-Down Text.**

Creating an expanding link

Shortcut	Tool Strip and Ribbon
Alt+N, P, T	Insert > Expanding Text

To create an expanding link:

1 Open the topic that will contain the expanding text link.

2 Highlight the expanding link.
The ⊥T expanding text link icon appears after the expanding text link.

3 Click inside the empty, second set of brackets.

4 Type the expanding link text.

Creating a toggler link

Shortcut	Tool Strip and Ribbon
Alt+N, O	Insert > Toggler

To create a toggler link:

1 Open the topic that will contain the toggler link.

2 Right-click the toggler content's block in the block bar and select **Name.**

3 Type a name for the toggled element.

4 Click **OK.**

5 If needed, assign the same name to other content blocks.

6 In the topic, highlight the text that you want to use as the toggler link.

7 Select **Insert** > **Toggler**.

The Insert Toggler dialog box appears.

🔴 Insert Toggler	? ✕

The hotspot text: [Click Me!]

Toggler class: | Toggler targets:

.footer		Named Elements	Tag	T...
		☐ MyTogglerContent	p	D...
.footnote-text				
.Topic-Text-Onestep				
.Topic-Text-Numbered ⌄		Go To

| OK | Cancel |

8 Select a toggler target by checking its checkbox.

Remember, you can associate more than one toggler target with a toggler link.

9 Click **OK**.

The ⎕ toggler icon appears to the left of the toggler text.

Related topic, keyword, and concept links

You can add related topic, keyword, and concept (also known as "see also") links to your topics. When the user clicks one of these links, a popup window appears with a list of topics:

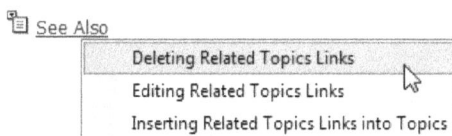

    Related topic, keyword, and concept links can only be used in online and EPUB targets. If you are creating an EPUB target, you can enable or disable dynamic content such as related topic links. See 'Setting up a print target' on page 338.

'What's the difference?'

All three of these links open a popup list of topics. The difference between them is how you select the topics that appear in the list.

Related topics links display a list of topics that you have manually selected. They are easier to create than keyword and concept links, but they are much harder to update. If you need to update a related topics link, you must manually add topics to or remove topics from the list.

Keyword links display a list of topics that include the specified index term(s) (or "keywords"). If you remove a keyword from or add a keyword to a topic, all of the keyword links that use the keyword are automatically updated.

Concept links display a list of topics that include the same concept term. If you remove a concept term from or add a concept term to a topic, all of the concept links that use that concept term are automatically updated.

Creating a related topics link

Shortcut	Tool Strip and Ribbon
Alt+N, R, T	Insert > Related Topics Control

To create a related topic link:

1 Open the topic that will contain the related topics link.

2 Position your cursor where you want to insert the related topics link.

3 Select **Insert > Related Topics Control**.
The Insert Related Topics Control dialog box appears.

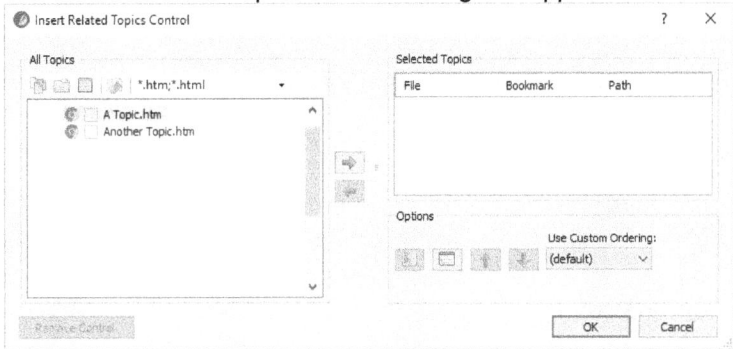

4 Select a topic.

5 Click ➡ to add the topic to the related topics link.

6 Add more topics as needed.

7 Click **OK**.
The related topics link appears in your topic.

Creating a keyword link

Before you create keyword links, you will need to add index keywords to your topics. When you create a keyword link, you select a keyword to include all of the topics that contain the keyword. See "Indexes" on page 195 for information about adding keywords to topics.

Shortcut	Tool Strip and Ribbon
Alt+N, Y	Insert > Keyword Link Control

To create a keyword link:

1 Open the topic that will contain the keyword link.

2 Position your cursor where you want to insert the keyword link.

3 Select **Insert** > **Keyword Link Control**.
The Insert Keyword Link Control dialog box appears.

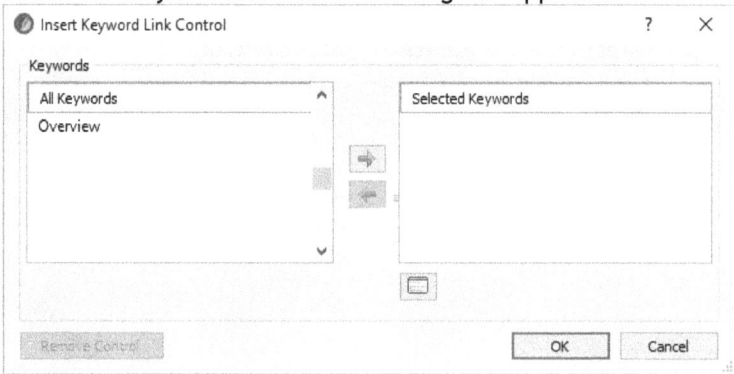

4 Select a keyword.

5 Click ⇨ to add the keyword to the keyword link.

6 Add more keywords as needed.

7 Click **OK**.
The keyword link appears in your topic.

Keyword links do not work in the preview.

Creating a concept link

Before you create a concept link, you need to add concept terms to your topics. When you create a concept link, you select a concept term to include all of the topics that contain the concept term.

Once advantage of concept links is that you can reuse the terms as search filters. For example, the Flare help system uses numerous concept terms, including "Best Practices" and "FAQs." When you search the help system, you can filter the results to only include topics that contain a selected concept term. For more information about setting up search filters, see "Adding search filters" on page 210.

Shortcut	Tool Strip and Ribbon
Alt+N, C	Insert > Concept Link

To add a concept term:

1 Open a topic to associate with the concept term.

2 Position the cursor where you want to add the concept term.

3 Select **View** > **Concept Window**.
The Concepts window appears.

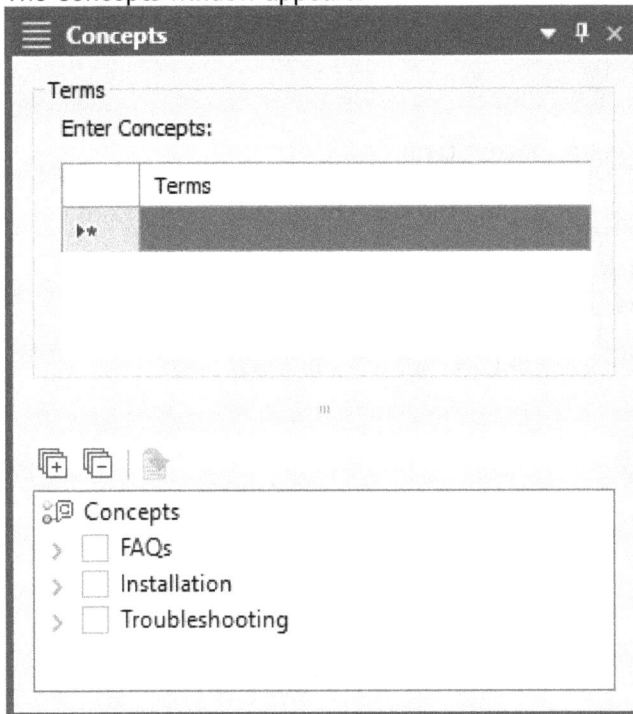

4 Type a concept term and press **Enter**.

5 Type more terms as needed.

6 Click **Save**.

To add a concept link:

1 Open the topic that will contain the concept link.

2 Position your cursor where you want to insert the concept link.

3 Select **Insert** > **Concept Link**.
The Insert Concept Link Control dialog box appears.

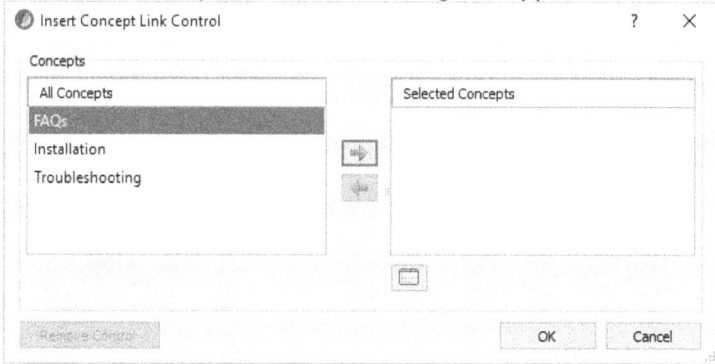

4 Select a concept term.

5 Click to add the concept term to the concept link.

6 Click **OK**.
The concept link is added to the topic.

Concept links do not work in the preview.

Relationship links

You can create relationship tables to organize your topics by type. For example, you can create a relationship table to specify how the following topics are related to each other:

Relationship	Concept	Task	Reference
soccer	defense.htm offense.htm	goalkeeping.htm passing.htm shooting.htm	rules.htm

This relationship table could be used to add the following links to your topics:

Related Information
- Defense
- Offense

Related Tasks
- Goalkeeping
- Passing
- Shooting

Reference Materials
- Rules

'How are relationship links different from help controls?'

There are four key differences between relationship links and help controls such as related topic, keyword, and concept links:

1 **How they are created and applied**
  Relationship links are created based on relationship tables, and you can associate different relationship tables with different targets.

  Help controls are created by manually selecting topics or by automatically selecting topics that contain selected keyword or concept markers.

2 **Link grouping**
Relationship tables are used to specify how topics are related based on topic types such as concept, tasks, and reference. Relationship links can separate links based on their type and display link group headings such as "Related Information," "Related Tasks," and "Reference Materials."

Help controls display the links in one list.

3 **Appearance**
Relationship links appear in the topic.

Help control links appear in a popup window.

4 **Print support**
Relationship links will appear in print targets.

Help control links do not appear in print targets.

Creating a relationship table

You can use one relationship table for your project, or you can create multiple relationship tables and use them for different targets.

Shortcut	Tool Strip	Ribbon
Ctrl+T	(Content Explorer	File > New

To create a relationship table:

1 Select **File** > **New**.
The Add File dialog box appears.

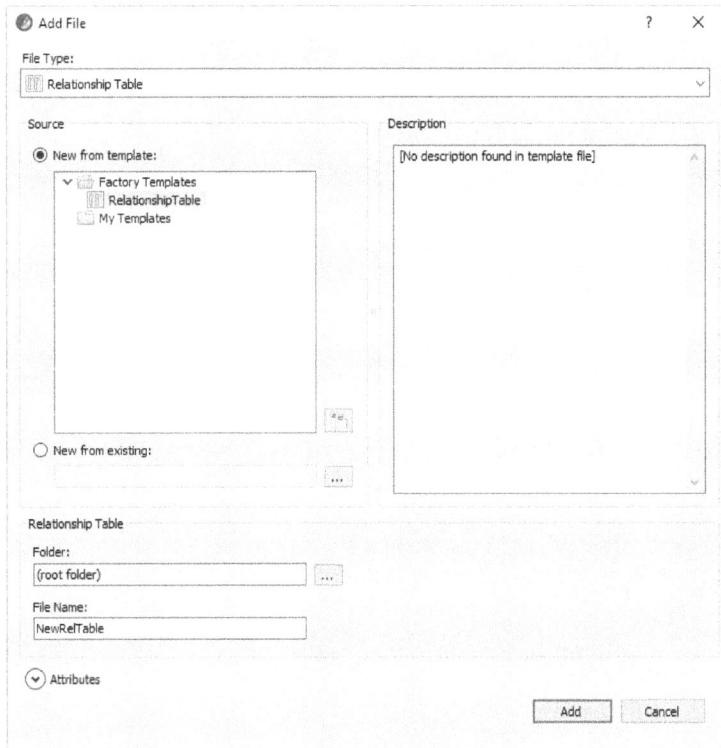

2 For **File Type**, select **Relationship Table**.

3 Select a **Source** template.

4 Type a **File Name**.

5 Click **Add**.

Adding a relationship to a relationship table

Relationships are used to group topics by type. For example, you can create a relationship to group five topics about printing to specify whether each topic is a conceptual, task, or reference topic.

To add a relationship to a relationship table:

1 Open a relationship table.

2 Click ▦ to create a new row.

3 Click 📝.

The Row Properties dialog box appears.

4 Type a name for the row.

5 Click **OK**.

6 Click the **Concept**, **Task**, or **Reference** cell.

7 Click 📄.

8 Select a topic and click **OK**.

Adding a column to a relationship table

By default, relationship tables include concept, task, and reference columns. You can add columns to specify other types of topic relationships.

To add a column to a relationship table:

1 Open a relationship table.

2 Click inside a column.

3 Click 📊.

4 Right-click inside the column and select **Column Properties**.
The Column Properties dialog box appears.

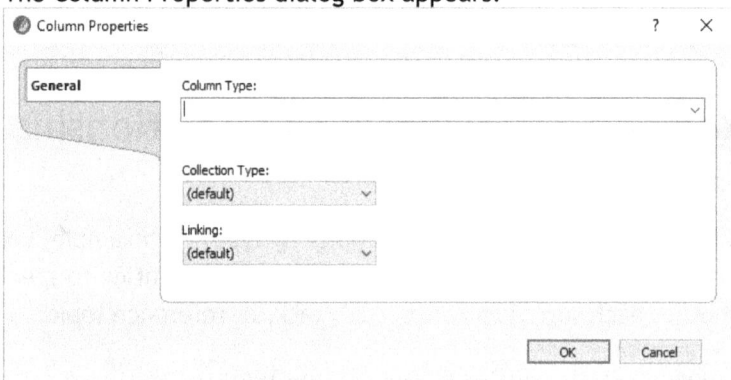

5 Type or select a **Column Type**.

6 Select a **Collection Type**.

Type	Description
Unordered	Creates an unordered list of links.
Family	Creates links to topics in the row and all of the topics in the same cell.
Sequence	Creates links based on their order in the relationship table. This option is only available for DITA topics.
Choice	Allows you to highlight a link in the group. This option is only available for DITA topics.
Use CONREF target	Uses the CONREF attribute determine the links. This option is only available for DITA topics.

7 Select a **Linking** option.

Option	Description
Source Only	The topic will link to other topics, but other topics will not link back to it.
Target Only	The topic will not link to other topics, but other topics will link to it.
None	The topic will not link to other topics, and other topics will not link to it.
Normal	The topic will link to other topics, and other topics will link to it.
Use CONREF target	Uses the CONREF attribute determine the links. This option is only available for DITA topics.

8 Click **OK**.

Renaming a column in a relationship table

You can rename the default "concept," "task," and "reference" columns or columns you have added to a relationship table.

To rename a column to a relationship table:

1 Open a relationship table.

2 Click inside a column.

3 Click 🔲.

The Column Properties dialog box appears.

4 Type a new **Column Type**.

5 Click **OK**.

Creating a relationship link

You can add relationship links to your topics to automatically insert links into your topics based on the defined relationships in a relationships table.

Shortcut	Tool Strip and Ribbon
Alt+I, Y, A	Insert > Proxy > Relationships Proxy

To create a relationship link:

1 Open the topic that will contain the relationship link.

2 Position your cursor where you want to insert the link.

3 Select **Insert > Proxy > Relationships Proxy**.

The Relationships Proxy dialog box appears.

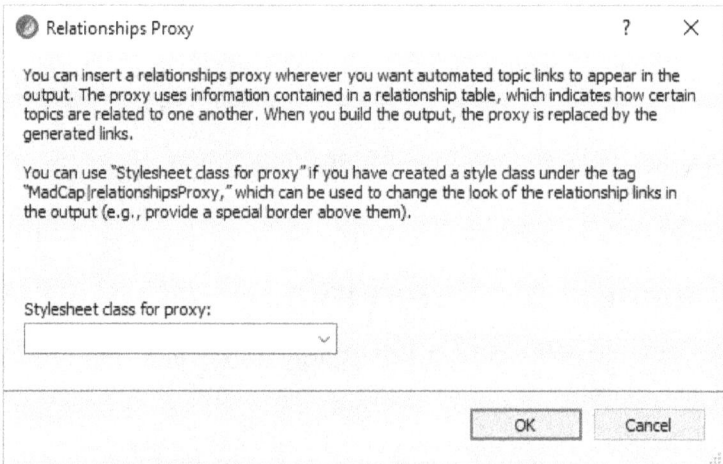

4 Click **OK**.

The relationships proxy appears in your topic.

Associating a relationship table with a target

You can create and use different relationship tables for each target to control which links appear in your topics.

To associate a relationship table with a target:

1 Open a target.

2 Select the **Relationship Table** tab.

3 Select the **Relationship Table(s)** to use.

4 Save the target.

Sample questions for this section

1 How can you view a list of topics that link to a topic?

A) Select **Topic > Show Links**.

B) Right-click the topic in the Content Explorer and select **View Links**.

C) Right-click the topic and select **Topic Properties**.

D) You can't.

2 How can you find and fix broken links?

A) Click each link in the XML Editor.

B) Select **Project > Check Links**.

C) Select **Analysis > Links > Broken Links**.

D) Select **Edit > Find and Replace** and search for broken links.

3 How can you view a list of topics that include an image?

A) Select **View > Image Links**.

B) Open the **Link List**.

C) Select **Report > Images > Link List**.

D) Right-click the image in the Content Explorer and select **View Links**.

4 How do you change the link label for a cross reference?

A) Change the **Link Label** in the Cross References dialog box.

B) Change the MadCap|xref style properties.

C) Right-click the cross reference and select **Edit Link Label**.

D) Just retype it in the topic.

5 Which type of link can open a web page? (select all that apply)

A) Hyperlink

B) Popup

C) Text popup

D) Cross reference

6 Which type of link(s) can *only* show and hide content directly below the link?

A) Expanding links

B) Drop-down links

C) Toggler links

D) Drop-down and toggler links

7 Which type of link displays a list of topics that contain a specified index marker?

A) Related topic

B) Keyword

C) Concept

D) See Also

8 Which type of link does not work in Flare's preview window?

A) Popup links

B) Cross references

C) Expanding links

D) Keyword links

Navigation

This section covers:

- TOCs
- Index
- Search
- Glossaries
- Browse sequences

TOCs

A table of contents ("TOC") is an ordered list of links that your users can use to find and open topics. Most TOCs start with introductory topics and end with troubleshooting and advanced topics.

A TOC contains books and pages. Books are used to organize pages and add levels to your TOC. They can link to topics, or they can simply be used to group pages. Pages always link to topics or other content. A page does not have to be inside a book. For example, a "What's New" page is often placed at the beginning of a TOC to draw the user's attention.

You don't have to include every topic in your TOC. If you don't include a topic in your TOC, it won't be included when you create a print target.

TOC books and pages usually link to topics, but they can also link to websites, email addresses, other TOCs, browse sequences, and documents such as .doc, .xls, and .pdf files.

TOCs are stored in an XML-based .fltoc file in the Project Organizer.

Moving the TOC to the accordion TIP▷

You can move the TOC Editor to the accordion if you want to use your TOC as an alternative to the Content Explorer.

TIP▷ *You can Ctrl-click a book or page in the TOC to open its associated topic in the XML Editor. If you click ⬚, you can double-click TOC items to open them in the XML Editor rather than opening the Properties dialog box.*

Shortcut	Tool Strip and Ribbon
Alt+W, D	Window > Dock

To dock the TOC Editor:

1　Open your TOC in the TOC Editor.

2　Select **Window** > **Dock**.
　Your TOC will move to the left dock.

Creating a TOC

When you create a project, Flare creates a blank TOC for you. You can use this TOC, or you can create your own.

'Can I create multiple TOCs?'

Yes, you can create multiple TOCs in a project. If you create multiple TOCs, you can use different TOCs for different targets. For example, you can include different topics and organize your topics in a different order for print and online targets. You can also create multiple TOCs if you have multiple authors.

Shortcut	Tool Strip	Ribbon
Ctrl+T	📄 (Content Explorer	File > New

To create a TOC:

1　Select **File** > **New**.
　—OR—
　Right-click the TOCs folder and select **Add Table of Contents**.
　The Add File dialog box appears.

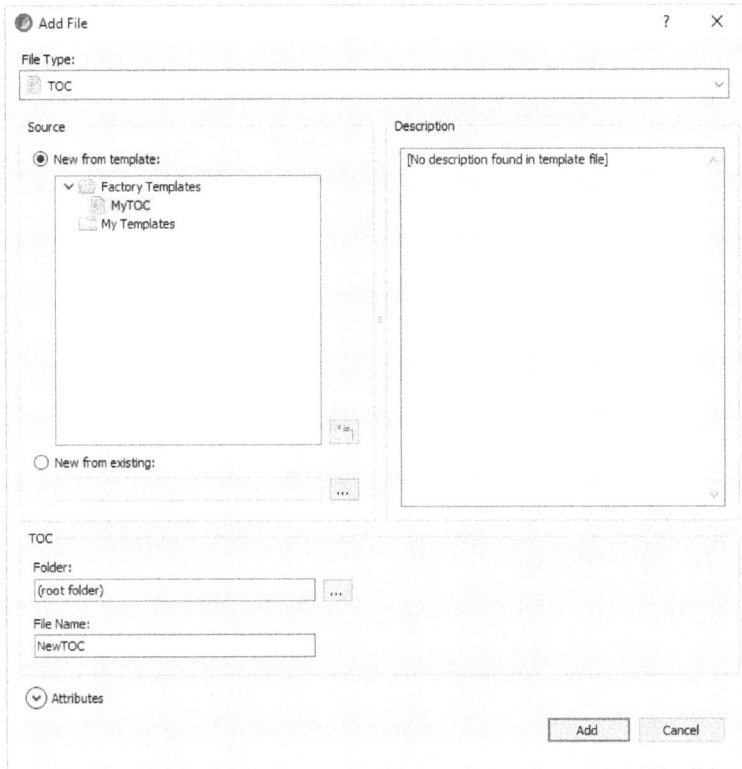

2 For **File Type**, select **TOC**.

3 Select a **Source** template.

4 Type a **File Name**.

5 Click **Add**.

The TOC appears in the TOC folder in the Project Organizer and opens in the TOC Editor.

Creating TOC books and pages

Flare provides multiple ways to create TOC books and pages. You can click the new TOC book or new page icons, type a title, and select a link. Or, you can drag a topic from the Content Explorer to the TOC. You can even drag a folder to the TOC and auto-create a book for the folder and pages for each topic inside the book.

To create a TOC book:

1 Open a TOC.

2 Click ▓ in the TOC Editor toolbar.
A new TOC book named **New TOC Book** appears.

3 Type a name for the book.
—OR—

Click ▓ to use a variable in the label.

4 If necessary, move the book to a new location in the TOC.
You can drag-and-drop the TOC book or use the arrows in the
TOC Editor toolbar.

To create a TOC page:

1 Open a TOC.

2 If you want to add a page inside a book, select the book.

3 Click ▓ in the TOC Editor toolbar.
A new TOC page named "New Entry" appears.

4 Double-click the TOC page.
—OR—

Click ▓ in the TOC Editor toolbar.

The Properties dialog box appears.

5 Type a **Label** for the page.

—OR—

Click to use a variable in the label.

> ▥➤ *If you use the "LinkedHeader" variable (in the System variable set), your TOC item's label will automatically match the h1 in the selected topic.*

6 Click **Select link**.

The Link to Topic dialog box appears.

7 Select a topic.

8 Click **Open**.

9 Click **OK**.

> ◇ *A flag (✐) icon appears beside TOC books and pages that are not linked. A broken link (✐✗) icon appears beside books and pages that have broken links.*

Finding and fixing issues in a TOC

You can find and fix TOC books and pages that are not linked or have broken links.

To find and fix TOC issues:

1 Open a TOC.

2 If your TOC books are intentionally unlinked, click ▥.

3 Click ⚓.

The first issue is highlighted. Unlinked items are marked with a ✐ icon. Broken links are marked with a ✐✗ icon.

4 Right-click the TOC item and select **Properties**.

The Properties dialog box appears.

5 On the **General** tab, select a new link and click **OK**.

'Stitching' a PDF into a TOC

You can "stich" a PDF into a TOC to include the PDF when you build a target. In online targets, the PDF will be included with the other output files and open in Acrobat reader or the browser window, depending on the user's browser. In PDF targets, the PDF will be included in the output PDF.

A stitched PDF cannot link to any other content in the target, and other content cannot link into the stitched PDF.

To 'stitch' a PDF into a TOC:

1   Open a TOC.

2   Drag the PDF from the Content Explorer or File List into the TOC.

Auto-generating TOC books and pages

If you have a long topic with multiple subheadings, you can auto-generate TOC pages to link to the subheadings. For example, the following topic has one main heading and four sub-headings:

South Africa (formatted as Heading 1)
  Cape Town (Heading 2)
  Durban (Heading 2)
  Johannesburg (Heading 2)
  Pretoria (Heading 2)

You can auto-create the following TOC entries for this topic:

 South Africa

  Cape Town
  Durban
  Johannesburg
  Pretoria

To auto-generate TOC entries:

1 Double-click a TOC page.
The Properties dialog box appears.

2 Select the **Auto-generate** tab.

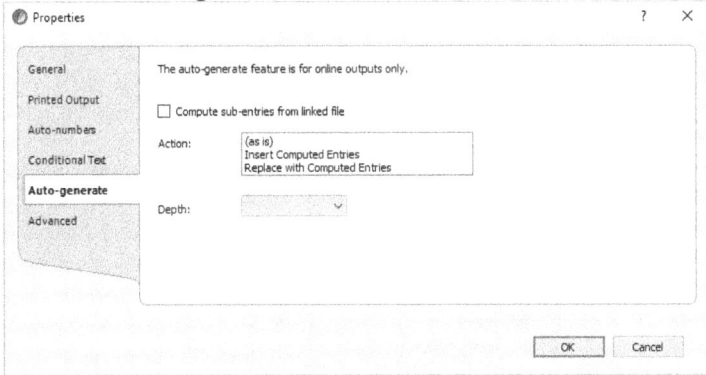

Properties	? ×

General

Printed Output

Auto-numbers

Conditional Text

Auto-generate

Advanced

The auto-generate feature is for online outputs only.

☐ Compute sub-entries from linked file

Action:
(as is)
Insert Computed Entries
Replace with Computed Entries

Depth:

OK   Cancel

3 Select **Compute sub-entries from linked file**.

4 Select an **Action**.

☐ **Insert Computed Entries** adds the TOC entries below the selected TOC entry.

☐ **Replace with Computed Entries** replaces the selected TOC entry with the new entries.

5 Select a heading **Depth**.
The heading depth is the level of subheadings that should be automatically included in the TOC. For example, selecting **2** will include pages for all heading levels 1 and 2.

6 Click **OK**.

Linking TOCs

If you create multiple TOCs, you can create a link from one TOC to another TOC.

To link TOCs:

1 Open a TOC.

2 Select the location in the TOC where you want to add the link to the other TOC.

3 Click ☐.
A new TOC entry named "New entry" appears.

4 Select the new entry and click 📄.
The Properties dialog box appears.

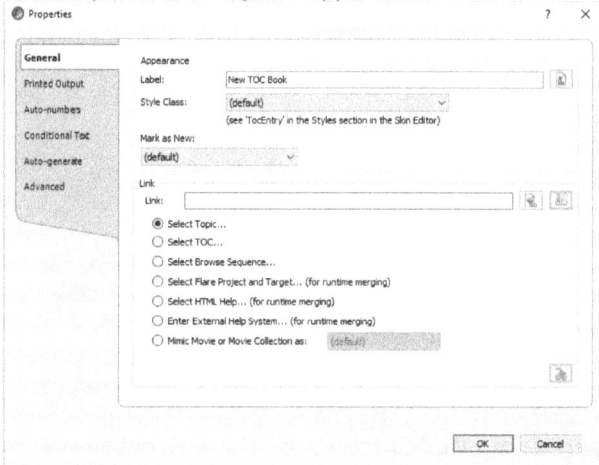

5 Click **Select TOC**.
The Link to TOC dialog box appears.

6 Select the TOC to which you want to link the page.

7 Click **Open**.

8 Click **OK**.
The icon in the TOC Editor changes to 📄, indicating that the page is linked to a TOC.

Deleting a TOC book or page

You can delete books or pages from your TOC. If you delete a book or page, Flare does not delete the topic to which it is linked.

To delete a TOC book or page:

1 Open a TOC.

2 Select a book or page.

3 Press **Delete**.

The book or page is removed from your TOC.

Applying a TOC to all targets

If you have multiple TOCs, you can select a TOC to be used when you build any target. If you only have one TOC, Flare will automatically associate it with your targets.

To apply a TOC to all targets:

1 Select **Project** > **Project Properties**.

The Project Properties dialog box appears.

2 Select the **Defaults** tab.

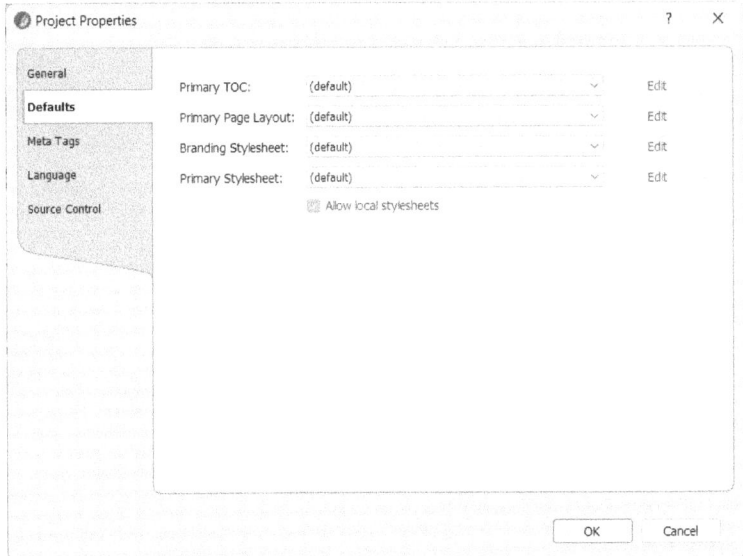

3 Select a **Primary TOC**.

4 Click **OK**.

Applying a TOC to a target

If you have multiple TOCs, you can select a TOC to be used when you build a target. If you only have one TOC, Flare will automatically associate it with your targets.

To apply a TOC to a target:

1 Open a target.

2 On the **General** tab, select a **Primary TOC**.

Specifying chapter breaks

If you add chapter breaks to your TOC, you can create separate Word, FrameMaker, or PDF documents for each chapter.

To specify a chapter break:

1 Open your TOC.

2 Right-click a TOC book or page and select **Properties**. The Properties dialog box appears.

3 Select the **Printed Output** tab.

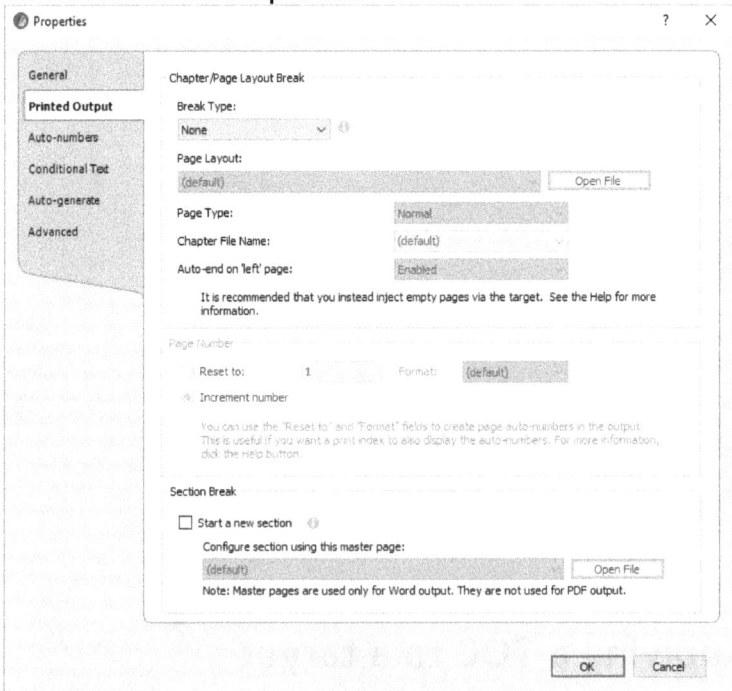

4 For **Break Type**, select **Chapter Break**.

5 Click **OK**.

Finding topics that are not in the TOC

You can use the Topics Not in Selected TOC report to find topics that are not in a TOC. You don't have to include a topic in a TOC, but this report can help you find topics you may have forgotten to include.

To find topics that are not in a TOC:

1 Select **Analysis > More Reports > Topics Not In Selected TOC**.

2 For **Filter**, select a TOC.

TIP *You can double-click a topic in the list to open it in the XML Editor.*

Finding and opening topics from the TOC

When you hover over a TOC book or page, a tooltip will appear that includes the path of TOC item's destination. However, you can also click the Locate file in Content Explorer button to quickly find the topic. If you often need to open topics from the TOC, you can set Flare to open topics when you double-click a TOC item.

To find a topic in the Content Explorer from the TOC:

☐ Right-click a TOC book or page and select **Locate in Explorer**.
   —OR—
   Select a TOC book or page and click ⬚.

To open topics from the TOC when you double-click a book or page:

☐ Click ⬚.

Finding a topic in the TOC

You can also locate a topic in the TOC. For example, you might move a topic to a new folder in the Content Explorer and decide to also change where it appears in your TOC.

To find a topic in the TOC:

☐ Right-click a topic and select **Locate in TOC**.
—OR—
Highlight a topic and select **Project** > **Locate in TOC**.

Adding a TOC to a print target

To include a table of contents in a print target, you need to either create a TOC topic or set your target to automatically add a TOC. If you create your own TOC topic, you can format the topic's heading and specify exactly where you want the TOC to appear in your target. If you set your target to automatically add a TOC, Flare will add it at the beginning of the print target or after the title page (if you have one).

To create a TOC topic:

1 Select **File** > **New**.
The Add File dialog box appears.

2 For **File Type**, select **Topic**.

3 For **Source**, select **New from Template** and select the **TopicForTOC** template.

4 Select a **Folder**.

5 Type a **File Name** for the TOC topic.

6 Click **Add**.
The TOC topic appears in the Content Explorer and opens in the XML Editor.

7 Add the new TOC topic to your TOC in the Project Organizer.

To set a print target to automatically include a TOC:

1 Open a print target.

2 Select the **Advanced** tab.

3 Select **Insert TOC proxy**.

Adding a TOC to an online target

To include a TOC in an HTML5 tripane or WebHelp online target, you will need to enable the TOC in the skin. In an HTML5 target that uses a topnav or sidenav skin, the TOC appears as a menu and is automatically enabled.

To include a TOC in an HTMl5 tripane or WebHelp skin:

1 Open a skin.

2 On the **General** tab, select the **TOC** option.

Indexes

Like FrameMaker, Flare adds your keywords as markers inside your topics. You can copy a keyword to other topics or delete a keyword when you delete its associated content.

There are four ways to add index entries to topics:

- ☐ "Quick term" method
- ☐ Index Window method
- ☐ Index Entry Mode method
- ☐ Auto-index phrase set method

'Why should I create an index?'

Most users will use a search tool instead of an index. An index also takes much more time to create than the search, since you must add index keywords to your topics. However, a good index is more useful that the search because it only lists the most relevant topics. Another key advantage of the index over the search is that the index is included in print targets.

Adding index entries using the 'quick term' method

The quick term method can be used to quickly add a term to the index while you are writing. Because it's so efficient, I often use the quick term method to add index terms.

Shortcut	Tool Strip	Ribbon
F10	Tools > Index > Insert Keyword	Insert > Keyword

To add a term using the quick term method:

1 Open a topic.

2 Click before the word (or highlight the phrase) that you want to insert as an index term.

3 Select **Insert** > **Keyword**.
   —OR—
   Press **F10**.
   The term is added to the index. If you have Show Markers enabled, the term will appear in a green box.

   TIP *To show markers, click the* *icon's down arrow in the XML Editor toolbar and select* **Show Markers**. *If you can't see the entire index entry, increase the marker width.*

Adding index entries using the Index window method

The Index window can be used to add single word, multiple word, and second-level index entries. The Index window shows all of the index terms within the current topic.

Shortcut	Tool Strip and Ribbon
F9	View > Index Window

To add an index entry using the Index window:

1 Open a topic.

2 Click before or on the word or phrase that you want to insert as an index term.

3 Select **View** > **Index Window**.
   —OR—
   Press **F9**.
   The Index window appears.

4 Type a term or phrase and press **Enter**.
   The term or phrase is added to the index.

TIP *To add a second-level entry, include a colon between the first- and second-level entries.*

Adding index entries using the Index Entry mode method

Index Entry mode is useful when you need to add multiple index entries. When you switch to Index Entry mode, the words you type become index entries rather than topic content. It's a great tool for indexers, since they can focus on indexing and not worry about accidentally changing the content in a topic.

To add an index entry using Index Entry mode:

1   Open a topic.

2   Click ⬚ in the XML Editor toolbar.

3   Position the cursor where you want to add the index term.

4   Type the term and press **Enter**.
    The Index Entry window appears, and the term is added to the index.

5   Continue typing terms as needed. When you are done, click ⬚ in the XML Editor.

Automatically adding index entries

Instead of adding keywords to topics, you can create an auto-index phrase set to automatically add keywords to your topics when you build a target. This method may cause your targets to build more slowly, but you can share the list of keywords across projects.

If you plan to link your content to a Word or FrameMaker document, you should use this approach because the keywords are added when you generate a target. If you add keywords using the other methods, they will be removed when you re-import the document.

Shortcut	Tool Strip	Ribbon
Ctrl+T	📑 (Content Explorer	File > New

To create an auto-index phrase set:

1 Select **File** > **New**.
The Add File dialog box appears.

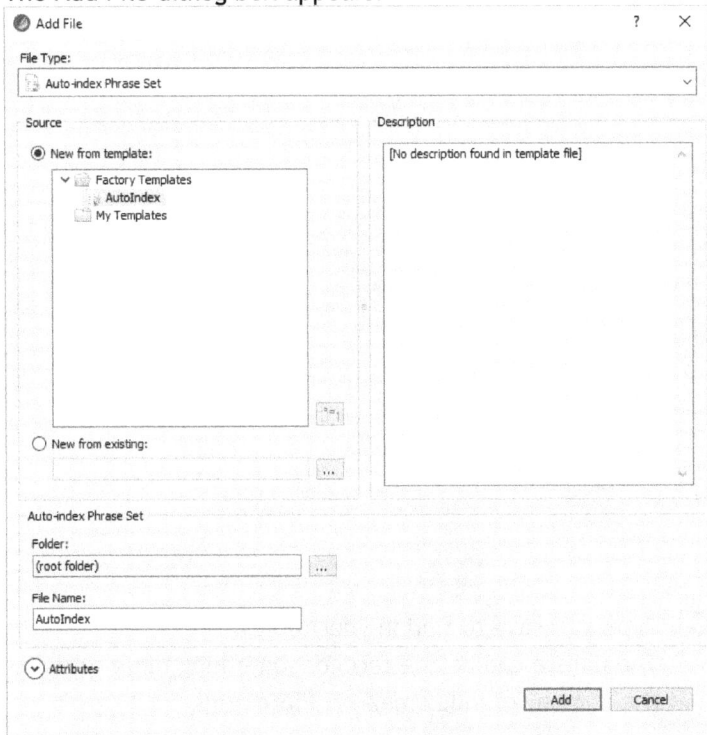

2 For **File Type**, select **Auto-index Phrase Set**.

3 Select a **Source** template.

4 Type a **File Name**.

5 Click **Add**.
The Auto-index Phrase Set is added to the Advanced folder in the Project Organizer and appears in the Auto-index Editor.

To add a term to an auto-index phrase set:

1   Open an auto-index phrase set.

2   Click 📑.

The Properties dialog box appears.

3   Type a **Phrase** to find in your topics.

The phrase can be the term or something more specific to limit the number of index markers Flare adds.

4   Type the **Index Term**.

5   Click **OK**.

The phrase and index term are case specific. If you want to add a keyword every time the word "pizza" or "Pizza" is used in your topics, you will need to add two auto-index entries.

Adding 'See also' index links

You can create *"See"* or *"See also"* index links to refer one index entry to another index entry.

Shortcut	Tool Strip	Ribbon
Ctrl+T	📑 (Content Explorer	File > New

To create an index link set:

1 Select **File** > **New**.
The Add File dialog box appears.

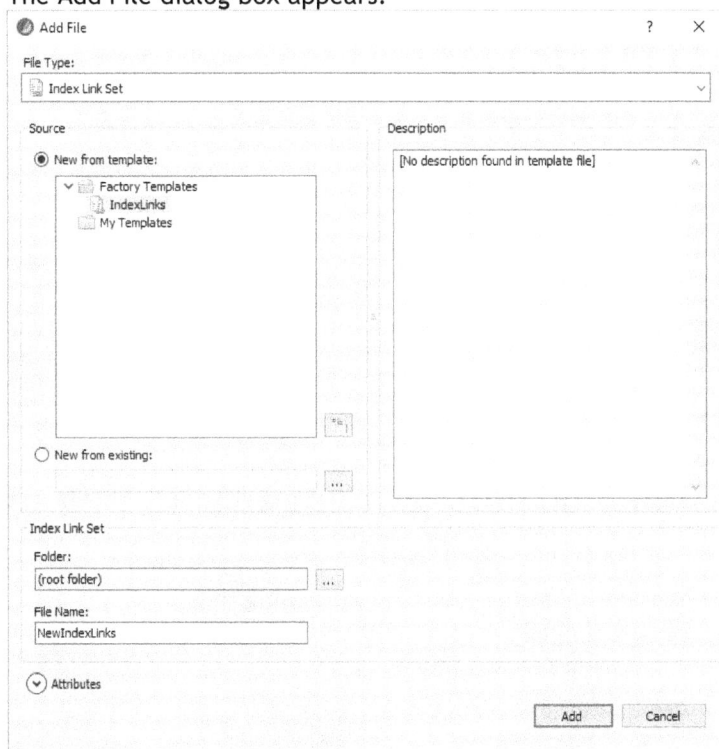

2 For **File Type**, select **Index Link Set**.

3 Select a **Source** template.

4 Type a **File Name**.

5 Click **Add**.
The Index Link Set is added to the Advanced folder in the Project Organizer and appears in the Index Links Editor.

To add a 'see also' entry to an index link set:

1 Open an index link set.

2 Click 📄.
The Properties dialog box appears.

3 Type an index **Term**.

4 Select a **Link** option.

 □ **See** – the "term" is not linked to any topics. The user should refer to the "linked term"

 □ **See Also** – the "term" is linked to topics. In addition, the user should also see the "linked term"

 □ **Sort As** – the "term" should be sorted as indicated by the 'linked term' (ex: ".HTML" sorted as "HTML")

5 Type the **Linked Term**.

6 Click **OK**.

Showing or hiding index entries in a topic

Index entry markers can be distracting when you are not indexing a topic. You can hide the index markers using the Show Tags icon in the XML Editor's toolbar.

To show or hide index markers:

1 Click the down arrow beside the 👁 ▾ icon.

2 Select **Show Markers**.
The index markers disappear.

 TIP▶ *You can change the size of the markers by modifying the Marker Width setting.*

Finding topics that are not in the index

You can use the Topics Not in Index report to find topics that do not contain index markers.

If you are using an auto-index phrase set, your topics do not contain markers and the report will list all of your topics.

To find topics that are not in the index:

☐ Select **Analysis > More Reports > Topics Not In Index**

You can double-click a topic in the list to open it in the XML Editor.

Reviewing index keyword suggestions

You can use the Index Keyword Suggestions report to review a list of suggested index keywords for your topics.

To review suggested index keywords:

1 Select **Analysis > Suggestions > Index Keyword Suggestions**.

2 To add a suggested keyword, highlight the item in the list and click **Apply**.

Viewing the index

Select **View > Index Window** to view the index. All of the index entries and listed at the bottom of the window.

If you are using an auto-index phrase set, your topics do not contain markers and the list will be empty.

Adding an index to a print target

To include an index in a print target, you need to either create an index topic or set your target to automatically add an index. If you create your own index topic, you can format the topic's heading and specify exactly where you want the index to appear in your target. If you set your target to automatically add an index, Flare will add it at the end of the print target.

To create an index topic:

1   Select **File** > **New**.
    The Add File dialog box appears.

2   For **File Type**, select **Topic**.

3   For **Source**, select **New from Template** and select the
    TopicForIndex template.

4   Select a **Folder**.

5   Type a **File Name** for the index topic.

6   Click **Add**.
    The index topic appears in the Content Explorer and opens in
    the XML Editor.

7   Add the new index topic to your TOC.

To set a print target to automatically include an index:

1   Open a print target.

2   Select the **Advanced** tab.

3   Select **Insert index proxy**.

Adding an index to an online target

To include an index in an online target (except HTML5 targets that use
a topnav skin), you will need to enable the index in the skin and
associate the skin with the target. To use an index in an HTML5 target
that uses a topnav skin, you will need to create an index topic and
include the topic in your TOC.

To include an index in a non-HTML5 topnav skin:

1   Open a skin.

2   On the **General** tab, select the **Index** option.

To include an index in an HTML5 topnav skin:

1 Select **File** > **New**.
The Add File dialog box appears.

2 For **File Type**, select **Topic**.

3 For **Source**, select **New from Template** and select the **TopicForIndex** template.

4 Select a **Folder**.

5 Type a **File Name** for the index topic.

6 Click **Add**.
The index topic appears in the Content Explorer and opens in the XML Editor.

7 Add the new index topic to your TOC.

Search

Most users use the search rather than the TOC or index to find information. When a user searches for a word or phrase, a list of all of the topics that contain the search term(s) appears. When the user opens a topic from the search, the search term is highlighted to make it easy to find.

Adding search synonyms

You can add search synonyms to include terms that do not appear in your topics. You can add two types of synonyms: directional and group.

Directional synonyms are used to associate specific terms with general terms. For example, a directional synonym might associate "country" with "New Zealand." If a user searches for "country," the search results would include topics that contain "country" or "New Zealand." However, if the user searches for "New Zealand," the results would only include topics that contain "New Zealand."

Group synonyms are used to associate a group of equivalent words. For example, a group synonym might include "close," "exit," and "quit." If a user searches for any of these terms, the results would include topics that contain any of these words.

Shortcut	Tool Strip	Ribbon
Ctrl+T	(Content Explorer	File > New

To add search synonyms:

1   Select **File** > **New**.
    The Add File dialog box appears.

2 For **File Type**, select **Synonyms File**.

3 Select a **Template Folder** and **Template**.

4 Type a **File Name**.

5 Click **Add**.

The synonym file is added to the Advanced folder in the Project Organizer and appears in the Synonym Editor.

To add a directional synonym:

1 Open a synonym file.

2 Select the **Directional** tab.

3 Click inside the **Word** cell beside the asterisk (*) and type the general term.

4 Click inside the **Synonym** cell and type the specific term.

5 If you want to also search for past tense and plural forms of the terms, select the **Stem** option.

To add a group synonym:

1 Open a synonym file.

2 Select the **Groups** tab.

3 Click inside the **Group** cell beside the asterisk (*) and type the terms separated by the = sign.
For example, **close=exit=quit**.

4 If you want to also search for past tense and plural forms of the terms, select the **Stem** option.

Excluding a topic in the full-text search

You can hide a topic in the full-text search. Many help authors use this feature to hide field-level context-sensitive help topics in the search.

To exclude a topic in the full-text search:

1 Right-click a topic in the Content Explorer or File List and select **Properties**.
—OR—
Select a topic and press **F4**.

2 Select the **Topic Properties** tab.

3 Deselect the **Include topic when full-text search database is generated** option.

4 Click **OK**.

Using search filters

You can use search filters to allow your users to search within a subset of your content. For example, you could add a search filter to allow user to only search installation or troubleshooting topics.

To add search filters, you will need to:

- [ ] Add concept terms to your topics (see "Creating a concept link" on page 170)

- [ ] Create a search filter set

- [ ] Use search filters in a target

To create a search filter set:

1 Select **File** > **New**.
   The Add File dialog box appears.

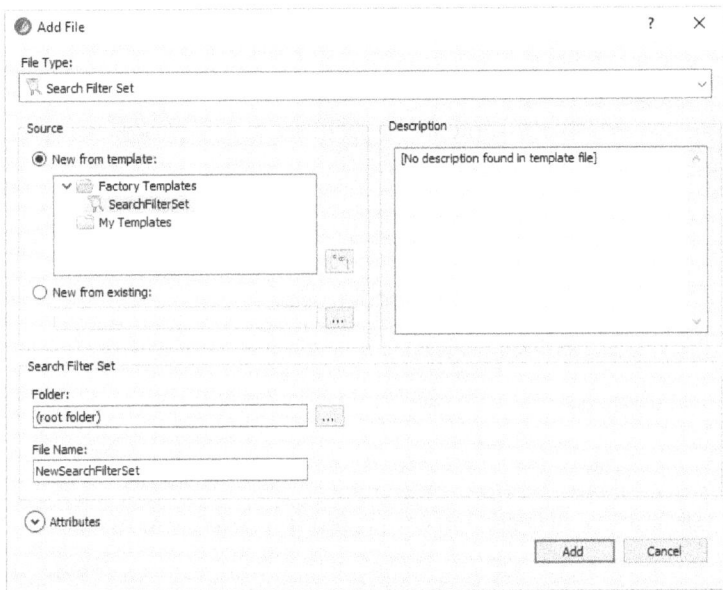

2 For **File Type**, select **Search Filter Set**.

3 Select a **Template Folder** and **Template**.

4 Type a **File Name**.

5 Click **Add**.
   The search filter set is added to the Advanced folder in the Project Organizer and appears in the SearchFilterSet Editor.

6 Double-click the "NewFilter" filter and type a new name.
   Examples: "FAQs," "Version 2," "Advanced Features"

7 Double-click the empty **Concepts** cell.
The Select Concepts for Search Filter dialog box appears.

Adding meta descriptions for SEO

A meta description is a brief summary of a topic. You can add meta descriptions to improve your content's results in search engines such as Google. Meta descriptions also appear in the HTML5 search results list. If you don't add meta descriptions, the content at the beginning of the topic will be used as the description.

◇ *Meta descriptions should be a maximum of 155 characters.*

To add a meta description to a topic (Flare 2021 and older):

1 Right-click a topic in the Content Explorer or File List and select **Properties**.
—OR—
Select a topic and press **F4**.

2 Select the **Topic Properties** tab.

3 Type a **Description**.

4 Click **OK**.

To add a meta description to a topic (Flare 2022 and later):

1 If needed, create a 'description' meta tag. See "Creating a meta tag" on page 401.

2 Right-click a topic in the Content Explorer or File List and select **Properties**.
—OR—
Select a topic and press **F4**.

3 Select the **Meta Tags** tab.

4 For a **Description** meta tag, type a **Value**.

5 Click **OK**.

Setting up featured snippets

Featured snippets display micro content blocks at the top of the search results. The micro content blocks are selected based on the user's search term(s), and they usually provide short answers to questions. In older versions of Flare, the featured snippet area could only include one micro content block. In Flare 2022 and later, you can include multiple micro content blocks as featured snippets.

To setup featured snippets for an HTML5 target:

1 Open an HTML5 target.

2 Select the **Search** tab.

3 Select a **Filter** to find micro content matches to the user's search term:

   □ all micro content files

   □ a specific micro content file

   □ micro content blocks that contain a specific meta tag

4 Select a maximum number of featured snippet results. The default is 1.

Formatting featured snippets

Featured snippets can be set to display as:

   □ **plain text** — the full micro content response appears

   □ **drop-down** — the micro content phrase is setup as a drop-down link, and the response appears/disappears when the link is clicked

   □ **truncated** — the micro content response is truncated if it is larger than a specified height

You can format the featured snippet area's background color and border; the link; and the link path. If you use the truncated option, you can specify the height of the featured snippet area and format the truncate button and area.

To format the featured snippets area:

1 Open an HTML5 skin.

   TIP *You can create a Search Results skin component to format the search results (including the featured snippet area).*

2 Select the **Setup** tab.

3 Select a **Featured Snippet View Mode**.

4 Select the **Styles** tab.

5 Open the **Featured Sippets** style group.

6 Set the properties as needed.

Setting up a knowledge panel

A knowledge panel displays micro content blocks in the search results that match the user's search term(s). The knowledge panel is typically added to the right of the search results, but it can also appear on the left or at the top of the search results (like featured snippets). Since it is larger than the featured snippet area, a knowledge panel often includes overview, summary, or procedural micro content blocks rather than the quick answer micro content blocks that typically appear as featured snippets.

To setup a knowledge panel for an HTML5 target:

1 Open an HTML5 target.

2 Select the **Search** tab.

3 Select a **Filter** to find micro content matches to the user's search term:

   ☐ all micro content files

   ☐ a specific micro content file

   ☐ micro content blocks that contain a specific meta tag

4 Select a maximum number of knowledge panel results. The default is 1.

Formatting the knowledge panel

Knowledge panels can be set to display as:

- ☐ **plain text** — the full micro content response appears

- ☐ **drop-down** — the micro content phrase is setup as a drop-down link, and the response appears/disappears when the link is clicked

- ☐ **truncated** — the micro content response is truncated if it is larger than a specified height

You can format the knowledge panel's background color, border, width, and position; the link; and the link path. If you use the truncated option, you can specify the height of the featured snippet area and format the truncate button and area.

To format the knowledge panel:

1 Open an HTML5 skin.

▣▶ *You can create a Search Results skin component to format the search results (including the knowledge panel).*

2 Select the **Setup** tab.

3 Select a **Knowledge Panel View Mode**.

4 Select the **Styles** tab.

5 Open the **Knowledge Panel** style group.

6 Set the properties as needed.

Glossaries

Flare provides two glossary features: a combined glossary tab/page that lists all of the terms and definitions, and glossary links that open definitions from your topics.

Glossary terms are stored in a glossary file with a .flglo extension. Glossary files are stored in the Glossaries subfolder in the Project folder.

Creating a glossary

You can use multiple glossaries in a project. For example, you can maintain a glossary of common terms that is shared across multiple projects and another glossary of project-specific terms.

Shortcut	Tool Strip	Ribbon
Ctrl+T	📄 (Content Explorer	File > New

To create a glossary:

1 Select **File** > **Add**.
   The Add File dialog box appears.

2 For **File Type**, select **Glossary**.

3 Select a **Source** template.

4 Type a **File Name**.
   Glossaries have a .flglo extension. If you don't type the extension, Flare will add it for you.

5 Click **Add**.
   The glossary appears in the Glossaries folder in the Project Organizer and opens in the Glossary Editor.

Adding a glossary term

You can add terms and definitions directly to a glossary or as you write topics. If you need to include formatting or an image in a definition, you can link to a topic for the definition.

1 Open a glossary.
   Glossaries are stored in the Glossaries folder in the Project Organizer.

2 Click 📄 in the Glossary Editor toolbar.
   The Properties dialog box appears.

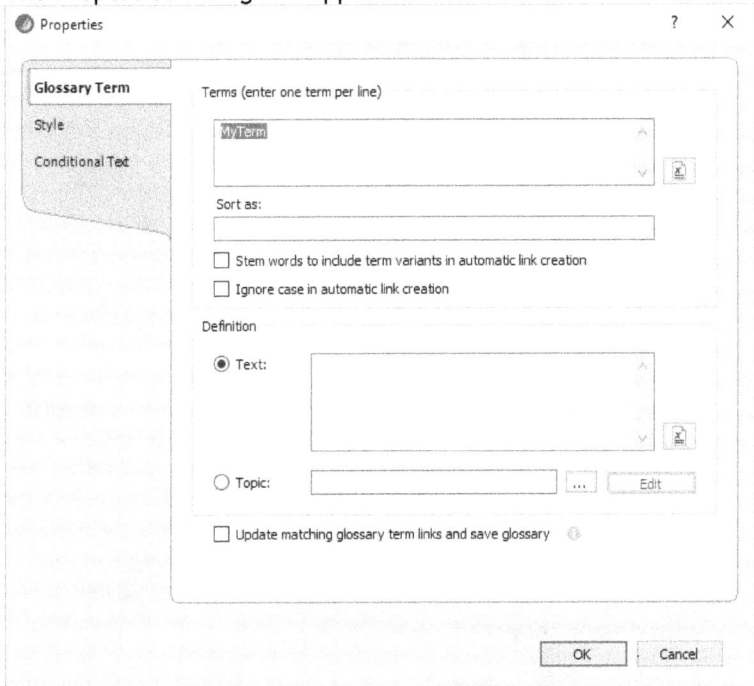

3 Type a glossary **Term**.

4 If you want to control how the term is sorted alphabetically, type a **Sort as** term/phrase. For example, .bmp could be sorted as "bmp" instead of ".bmp."

5 If you want to include variations of the phrase for glossary term popups/footnotes, select **Stem words to include term variants in automatic link creation**.

6 If you want to include case variations for glossary term popups/footnotes, select **Ignore case in automatic link creation.**

7 Type a **Definition** or select a topic that contains the definition.

8 Select the **Style** tab.

9 Select a style for the glossary link:

  □ **Expanding** — opens the definition with an expanding link

  □ **Popup** — opens the definition in a popup window

  □ **Hyperlink** — closes the current topic and opens the glossary page

  ◇ *If you don't select a style, Flare will use the "Popup" style.*

10 Click **OK.**

Adding a glossary to a print target

To include a glossary in a print target, you need to either create a glossary topic or set your target to automatically add a glossary. If you create your own glossary topic, you can format the topic's heading and specify exactly where you want the glossary to appear in your target. If you set your target to automatically add a glossary, Flare will add it at the end of the print target.

To create a glossary topic:

1 Select **File > New.**
The Add File dialog box appears.

2 For **File Type,** select **Topic.**

3 For **Source,** select **New from Template** and select the **TopicForGlossary** template.

4 Select a **Folder.**

5 Type a **File Name** for the glossary topic.

6 Click **Add.**

The glossary topic appears in the Content Explorer and opens in the XML Editor.

7 Add the new glossary topic to your TOC.

To set a print target to automatically include a glossary:

1 Open a print target.

2 Select the **Advanced** tab.

3 Select **Insert glossary proxy.**

To add glossary term footnotes to a print target:

1 Open a print target.

2 Select the **Glossary** tab.

3 Select a **Glossary Term Conversion** option.

4 Select the glossary or glossaries you want to use.

Adding a glossary to an online target

To include a glossary in an online target (except HTML5 targets that use a topnav skin), you will need to enable the glossary in the skin and associate the skin with the target. To use a glossary in an HTML5 target that uses a topnav skin, you will need to create a glossary topic and include the topic in your TOC.

To include a glossary in a non-HTML5 topnav skin:

1 Open a skin.

2 On the **General** tab, select the **Glossary** option.

To include a glossary in an HTML5 topnav skin:

1 Select **File** > **New**.
The Add File dialog box appears.

2 For **File Type**, select **Topic**.

3 For **Source**, select **New from Template** and select the **TopicForGlossary** template.

4 Select a **Folder**.

5 Type a **File Name** for the glossary topic.

6 Click **Add**.
The glossary topic appears in the Content Explorer and opens in the XML Editor.

7 Add the new glossary topic to your TOC.

To add glossary term popups to an online target:

1 Open a target.

2 Select the **Glossary** tab.

3 Select a **Glossary Term Conversion** option.

4 Select the glossary or glossaries you want to use.

'Where's the HTML Help glossary tab?'

HTML Help does not include a glossary tab, and MadCap cannot add one without requiring a .dll file. So, your glossary appears at the bottom of the TOC and opens on the right in the topic pane.

'Where's the HTML5 "topnav" or "sidenav" glossary?'

HTML5 "tripane" targets can include a Glossary tab in the navigation pane. HTML5 "topnav" targets do not have a navigation pane. If you want to include a list of all terms and definitions, you should create a glossary topic and include the topic in your target as described above.

Browse sequences

Like a TOC, a browse sequence is an ordered list of links that can be used to find and open topics. It even uses books and pages like a TOC.

You can create a browse sequence if you want to provide an alternate TOC for your users. For example, you could organize your TOC for technical support users and your browse sequence for managers. Or, you could organize your TOC by complexity (introductory topics first and troubleshooting topics last) and your browse sequence alphabetically.

Browse sequences are stored in a browse sequence file with a .flbrs extension. Browse sequence files are stored in the Advanced subfolder in the Project folder.

Creating a browse sequence

Shortcut	Tool Strip	Ribbon
Ctrl+T	(Content Explorer	File > New

To create a browse sequence:

1 Select **File** > **New**.
  The Add File dialog box appears.

2   For **File Type**, select **Browse Sequence**.

3   Select a **Source** template.

4   Type a **File Name** for the browse sequence.
    Browse sequences have a .flbrs extension. If you don't type the
    extension, Flare will add it for you.

5   Click **Add** and click **OK**.
    The browse sequence appears in the Advanced folder in the
    Project Organizer and opens in the Browse Sequence Editor.

To add books to a browse sequence:

1   Open a browse sequence.

2   Open the Content Explorer.

3   Drag a folder from the Content Explorer to the browse sequence.

4 If you need to rename the browse sequence book:

  ❑ Click the selected new book entry.
    —OR—
    Press **F2**.
    The text for the entry is highlighted.

  ❑ Type the new name.

To manually add books to a browse sequence:

1 Open a browse sequence.

2 Click 📖 or 📖 in the Browse Sequence Editor toolbar.
    A book named "New TOC Book" is added to the browse
    sequence.

3 Click the selected new book entry.
    —OR—
    Press **F2**.
    The text for the entry is now highlighted.

4 Type a name for the book.

To add pages to a browse sequence using drag-and-drop:

1 Open a browse sequence.

2 Open the Content Explorer.

3 Drag a topic from the Content Explorer to the browse
    sequence.

4 If necessary, use the arrows in the browse sequence toolbar to
    move the page up, down, left or right.

⬅ ➡ ⬆ ⬇

To add pages to a browse sequence using the Browse Sequence Editor:

1 Open a browse sequence.

2 Select the location in the browse sequence where you want to add the new entry.

3 Click ▤ in the Browse Sequence Editor.
An entry named "New entry" is added to the browse sequence.

4 Click the selected new entry.
—OR—
Press **F2**.
The text for the entry is now highlighted.

5 Type a name for the entry and press **Enter**.

6 If necessary, use the arrow buttons in the browse sequence toolbar to move the page left, right, up, or down.

⬅ ➡ ⬆ ⬇

7 Double-click the new entry.
The Properties dialog box appears.

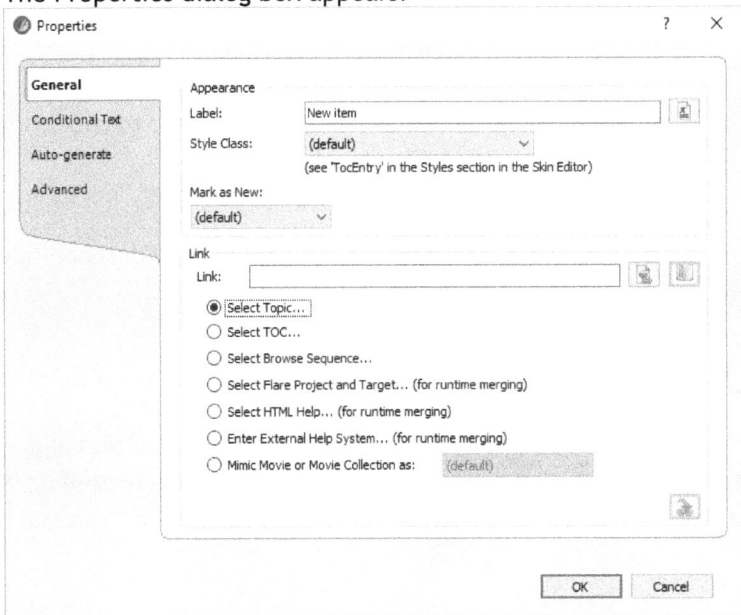

8 If needed, select the **Mark as New** option.
The page will be marked with the "New" icon: ⬚. You can change this icon in your skin.

9 Select **Select Topic**.

10 Click **Select Link**.
The Link to Topic dialog box appears.

11 Select a topic and click **Open**.

12 Click **OK**.

Creating a browse sequence based on your TOC 📝▷

Since Flare uses XML for the TOC and browse sequence files, you can make a copy of your TOC and convert it to a browse sequence.

To create a browse sequence based on your TOC:

1 In Windows Explorer, create a copy of your TOC file.
By default, TOC files are located in the Project\TOCs folder.

2 Paste the copy of your TOC file into the Project\Advanced folder.

3 Change the TOC file's extension from .fltoc to .flbrs.
The Rename dialog box appears.

4 Click **Yes**.
Your new browse sequence appears in the Advanced folder in the Project Organizer.

Using a browse sequence

To use a browse sequence, you need to enable it in a skin and associate it with a target. You can use different browse sequences in different targets, or you can use multiple browse sequences in the same target.

To enable a browse sequence in a non-HTML5 topnav skin:

1 Open a skin.

2 On the **General** tab, select the **Browse Sequence** option.

3 If you want to include "next" and "previous" browse sequence buttons in the WebHelp or Topic toolbar:

  ❑ Select the **WebHelp Toolbar** or **Topic Toolbar** tab.

  ❑ Add the **NextTopic**, **PreviousTopic**, and/or **CurrentTopicIndex** items to your toolbar.

The three items appear as follows (previous, current, and next):

To associate a browse sequence with a target:

1 Open a target.

2 On the **General** tab, select a **Browse Sequence**.

'Where are my HTML Help browse sequences?'

HTML Help does not include a separate browse sequence feature. Some tools, such as RoboHelp, use a .dll file to add browse sequences to HTML Help. MadCap does not want to require a .dll file for HTML Help, so Flare adds your browse sequence to the bottom of your TOC.

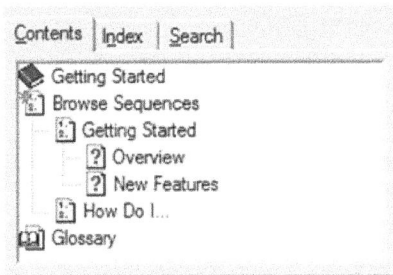

Sample questions for this section

1   Where are TOC files stored?
   A) In the Content Explorer's TOCs folder
   B) In the Content Explorer's Resources\TOCs folder
   C) In the Project Organizer's TOCs folder
   D) In the TOC Explorer

2   Which statement about a TOC is NOT true?
   A) Pages must be inside books.
   B) Pages can link to topics, Word documents, and web sites.
   C) Books can link to topics.
   D) Books can contain books.

3   How do you stitch an existing PDF into a PDF target?
   A) Select the PDF in your PDF target on the PDF Options tab.
   B) Drag the PDF into your TOC.
   C) Click **Stitch PDF** in the Project Properties dialog box.
   D) You cannot.

4   Which character is used to separate first- and second-level index entries?
   A) |
   B) /
   C) :
   D) ;

5   How can you view a list of your index terms?
   A) Double-click your index in the Project Organizer.
   B) Double-click your index in the Content Explorer.
   C) Select **View > Index Window**.
   D) You cannot view the index unless you build a target.

6   How do you exclude a topic in the search?
   A) Add the topic to your search file in the Project Organizer.
   B) Open the Topic Properties dialog box and deselect the **Include topic when full-text search database is generated** option.
   C) Do not include the topic in your TOC.
   D) You cannot exclude a topic from the search.

7 How do you add glossary term popups to an online target?
A) Select "Add popup definition" in the Properties dialog box.
B) Highlight the term in the topic and create a glossary hyperlink.
C) Select a **Glossary Term Conversion** option in your online target.
D) Glossary terms do not appear in online targets.

8 What is a browse sequence?
A) The path the user used to open the topic.
B) An ordered list of links that can be used to find and open topics, like a TOC.
C) A list of related topics at the bottom of topics.
D) A path at the top of your topics that shows how to find this topic in the TOC.

Format and design

This section covers:

- Stylesheets
- Branding stylesheets **NEW!**
- Page layouts
- Responsive layouts
- Template pages
- Skins
- Responsive output

Stylesheets

Like Word templates and FrameMaker catalogs, stylesheets are used to format content. You can define the formatting for each style in a stylesheet, and you can add your own styles (called "classes") to a stylesheet. You can even specify print- and online-specific styles within a stylesheet to use different formatting for print and online targets.

Stylesheets are stored in the Content Explorer. By default, stylesheets created in Flare are stored in the Resources\Stylesheets folder. If needed, you can move a stylesheet to a different folder.

External and inline styles

Styles that are defined in your cascading stylesheet are called "external" styles because the formatting information is stored in another file rather than in your topics.

You can also format your content directly by highlighting content and changing its appearance. For example, you can highlight a word and make it bold and red. This type of formatting is called "inline" formatting because the formatting information is stored directly in the topic.

You should avoid using inline formatting. It is much harder to change inline formatting than it is to change a style. If you change a style, all of the topics that use the style are automatically updated when you save the stylesheet. If you need to change inline formatting, you have to change it by hand. Another problem is that inline formatting overrides your styles. It is hard to maintain consistent formatting if you have scattered inline formatting throughout your topics.

If you have inline formatting, you can remove it or convert it to a style. See "Removing inline formatting from a table" on page 127, "Removing inline formatting" on page 246, and "Converting inline formatting to a style" on page 237 for more information.

Creating a stylesheet

Although most Flare projects use the same stylesheet for all topics, you can create multiple stylesheets and apply them to different topics. For example, you could create a "New Feature" stylesheet so that new topics in your project stand out to your users.

Shortcut	Tool Strip	Ribbon
Ctrl+T	📄 (Content Explorer	File > New

To create a stylesheet:

1 Select **File** > **New**.
   The Add File dialog box appears.

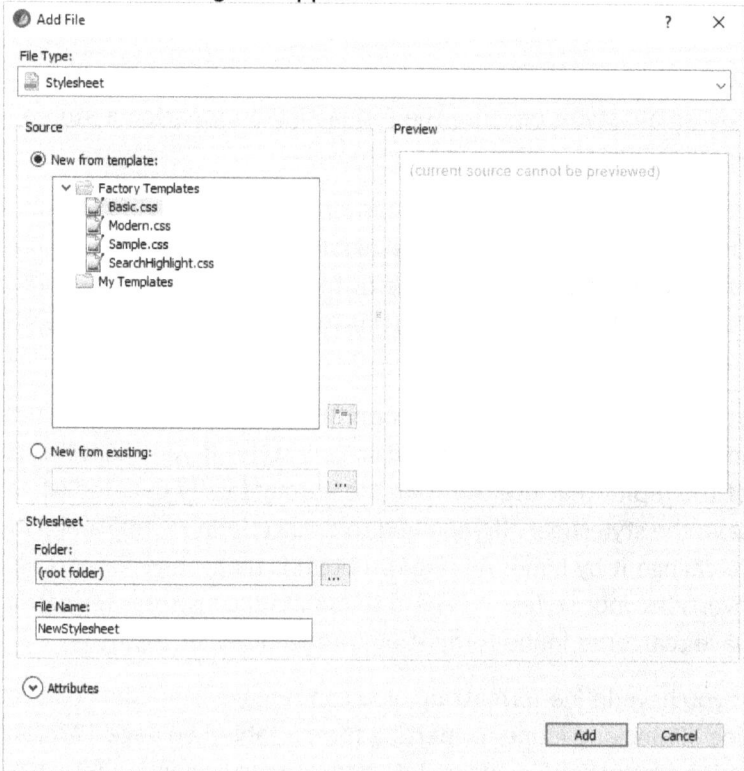

2 For **File Type**, select **Stylesheet**.

3 Select a **Source** template.

4 Type a **File Name** for the stylesheet.
Stylesheets have a .css extension. If you don't type the extension, Flare will add it for you.

5 Click **Add**.
The stylesheet appears in the Content Explorer and opens in the Stylesheet Editor.

Creating a style class

You can create style classes or ids to format notes, warnings, or other types of content. Most style classes are based on the p (paragraph) tag, but you can also create classes for headings, lists, and tables.

To create a style class:

1 Open a stylesheet.
The Stylesheet Editor appears.

2 Click **Add Selector**.

3 Select an **HTML Element**.
For example, to create a paragraph style class, select **p**.

4 Type a **Class Name**.

5 Click **OK**.

Creating an auto-numbering style class

You can create an auto-numbering style class to automatically number content, including headings, captions, and figures. Auto-numbering styles are often used in print documents.

To create an auto-numbering style class:

1 Open a stylesheet.
The Stylesheet Editor appears.

2 Click **Add Selector**.

3 Select an **HTML Element**.
For example, to create a paragraph style class, select **p**.

4 Type a **Class Name**.

5 Click **OK**.

6 Select the new style class.

7 If you are using the Simplified View:

- ☐ Double-click the style class.
- ☐ Select the **Auto-number** tab.

If you are using the Advanced View:

- ☐ In the Show drop-down box, select **Show: All Properties**.
- ☐ Open the **AutoNumber** property group.
- ☐ Click ⋯ for the **mc-auto-number** property. The Auto-Number Format dialog box appears.

8 Type or select an auto-numbering format.

Example	Auto-number Format
1.0 (sample text) 2.0 (sample text)	A:{n+}.{ =0}
1.0 　　1.1 　　1.2	A:{n}.{n+}
I. II.	B:{R+}.{ =0}
I. 　　A. 　　B.	B:{ }{A+}.
Chapter 1	CH:Chapter {chapnum}
Figure 1-1	CF: Figure {chapnum}-{n+}

9 Click **OK**.

10 Click **Save**.

Creating a redacted content style class

You can create a redacted content style class to "black out" sensitive, confidential, or private content in a PDF target. When you apply a redacted style to text, images, or other content, it is permanently replaced with a black rectangle. Users cannot view the original content.

TIP▶ *You can also use a redacted content style to highlight (rather than black out) content in a PDF target.*

To create a redacted content style class:

1 Open a stylesheet.
   The Stylesheet Editor appears.

2 Click **Add Selector**.

3 Select an **HTML Element**.
   For example, to create a paragraph style class, select **p**.

4 Type a **Class Name**.

5 Click **OK**.

6 Select the new style class.

7 If you are using the Simplified View:

   □ Double-click the style class.

   □ Select the **Advanced** tab.

   If you are using the Advanced View:

   □ In the Show drop-down box, select **Show: All Properties**.

   □ Open the **Redaction** property group.

   □ Click ⋯ for the **mc-redacted** property.

8 Select **Redacted**.

9 Click **Save**.

To select how redacted content appears in a PDF target:

1   Open a target.
The Target Editor appears.

2   Select the **Advanced** tab.

3   Select a **Redacted Text** option: blackout, highlighted, or normal.

Creating a 'Note' style class

You can create a style class that automatically adds text, such as the word "Note:" before your content.

To create a 'Note' style class:

1   Open a stylesheet in the Advanced View.
The Stylesheet Editor appears.

2   Click **Add Selector**.

3   Select **p** for the **HTML Element**.

4   Click **Advanced Options**.

5   Set the **Pseudo Element** to **Before**.

6   Click **OK**.

7   Select the **before** pseudo element.

8   Click ⋯ for the **content** property.

9   Type your content.
For example, type Note:

Selecting a color NEW!

You can select an RGB or CMYK color for text colors or background colors to customize styles, page layouts, and skins. Flare maintains a list of recently used colors, and you can even save colors that you use often. In Flare 2023 and later, you can select a CSS variable to use as a color value.

To select a color for a style:

1 Open a stylesheet.
The Stylesheet Editor appears.

2 If you are using the Simplified View:

☐ Double-click a style.

☐ Select the **color** or **background** property.

If you are using the Advanced View:

☐ Click ⋯ for the **color** or **background-color** property.

3 Select **More Colors**.
The Color Picker dialog box appears.

4 Select one of the following:

☐ ⬚ to select a color from your screen.

☐ **Recent** to select a recently used color.

- ☐ **Variables** to use a CSS Variable (see "Creating a CSS Variable" on page 248).

- ☐ **Saved** to select a saved color.

- ☐ **Named** to select a named color as defined by the W3C.

- ☐ **Web** to select a web-safe color.

5 If you want to use a CMYK color, select **Enable CMYK**.

TIP *If you select a CMYK color, you can convert it to RGB for print targets. See "Setting up a print target" on page 338.*

6 If you want to save your color, click .

7 Click **OK**.

Creating an image thumbnail style class

You can create a thumbnail style class to display smaller "thumbnail" versions of your images in your targets, and you can specify whether the image displays as full size when the user clicks or moves their mouse over an image.

*You can click and select **Show all Images as Thumbnails** to view all images as thumbnails in the XML Editor.*

To create an image thumbnail style class:

1 Open a stylesheet.
The Stylesheet Editor appears.

2 Select the **img** style or an img style class.

3 If you are using the Simplified View:

- ☐ Double-click the style class.

- ☐ Select the **Thumbnail** tab.

- ☐ Select a **Thumbnail** value:

  - ▪ **hover** — the image appears in a popup window when the user hovers the mouse over the thumbnail

- **link** — the image appears in a new window when the user clicks the thumbnail

- **popup** - the image appears in a popup window when the user clicks the thumbnail

□ Select a **Max Width** or **Max Height**.
The default setting is a height of 48px. Flare will proportionately scale the image, so you don't need to set both properties.

□ Click **OK**.

If you are using the Advanced View:

□ In the Show group, select Show: **All Properties**.

□ Open the **Thumbnail** property group.

□ Select a value for the **mc-thumbnail** property.

□ Select a value for the **mc-thumbnail-max-height** or **mc-thumbnail-max-width** property.
The default setting is a height of 48px.

□ Click **Save**.

Adding curved borders

You can add curved borders to tables or to any type of content, such as a "note" paragraph style. Microsoft Word and browsers that don't support curved borders will display square corners, but they are supported in most browsers and in PDFs.

To add curved borders:

1 Open a stylesheet in the Advanced View.
The Stylesheet Editor appears.

2 Select the style you want to edit.

3 Open the **Custom** property group.

4 Set the **border-radius** property.

Adding page breaks

You can use a style to add or avoid page breaks after, before, or inside a block of content. For example, you can set the h1 style to add a page break and always start on a new page. Or, you could set the ol and/or ul tags to avoid page breaks inside numbered and/or bulleted lists.

TIP▷ *The "avoid" setting specifies that there should not be a page break inside a block of content unless the entire block cannot fit on the page and must contain a page break.*

You can also add "independent" page breaks that don't use a style. Independent page break element can be used to specify special case page breaks that can be included or excluded in print targets using condition tags.

To add page breaks using a style:

1 Open a stylesheet in the Advanced View.
The Stylesheet Editor appears.

2 Select the style you want to edit.

3 Open the **PrintSupport** property group.

4 Set the **page-break-after**, **page-break-before**, or **page-break-inside** property.

To add an independent page break:

1 Open a topic or snippet.

2 Position your cursor where you want to insert the page break.

3 Select **Insert** > **Page Break**.
The page break appears in the topic or snippet.

Absolutely positioning content

You can absolutely position an element, such as an image or table, and flow content around it. Absolute positioning is supported in online and PDF targets.

To absolutely position content:

1 Open a stylesheet in the Advanced View.
The Stylesheet Editor appears.

2 Select the style you want to absolutely position, such as img.

3 Add a class:

- Click **Add Selector**.

- Type a **Class Name**.

- Click **OK**.

4 Select the style class.

5 Open the **Positioning** property group.

6 Set the **position** property to **absolute**.

7 Set the **top** property to specify the distance from the top of the browser or PDF viewer window.

8 Set the **left** property to specify the distance from the left of the browser or PDF viewer window.

9 Set the **z-index** property to one of the following values:

- **0** — to wrap text around the element (for PDF targets) or on the left (for online targets)

- **-1** — position the element behind the other content

- **1** — position the element on top of the other content

10 Open the **Box** property group.

11 Set the **margin** properties to specify the space around the element.

Formatting an iframe

You can use styles to specify an iframe's border, width, and/or height.

To format an iframe:

1   Open a stylesheet in the Advanced View.
    The Stylesheet Editor appears.

2   Select the **iframe** style.

3   Select and set a style property, such as border, height, or
    width.

Applying a condition tag with a style

You can set a style to automatically apply a condition tag to content.
For example, you could create a class of the img tag for screenshots
and set it to apply a "PrintOnly" condition tag. Any images that use the
class could then be excluded from online targets.

To apply a condition tag with a style:

1   Open a stylesheet in the Advanced View.
    The Stylesheet Editor appears.

2   Select a style.

3   In the Show drop-down box, select **Show: All Properties**.

4   Open the **Unclassified** property group.

5   Select a value for the **mc-conditions** property.
    Any content that uses the selected style will now also be
    tagged with the selected condition tag.

Converting inline formatting to a style

You can create a style class based on inline formatting. This approach
is useful for removing inline formatting or if you prefer to see your
formatting as you design it. You can create a style class in the
Stylesheet Editor, but sometimes it's easier to create a style by seeing
how it will look in your topics.

To create a style based on inline formatting:

1   Open a topic in which you want to use the new style.

2 Use the formatting options on the **Home** ribbon to format the text.

3 Click inside the formatted content.
◇ *If you want to create a character style, highlight the formatted content. If you want to make a paragraph style, do not highlight the content.*

4 Select **Home** > **Style Window**.
—OR—
Press **F12**.
The Style window appears.

5 Click **Create Style**.
The Create Style dialog box appears.

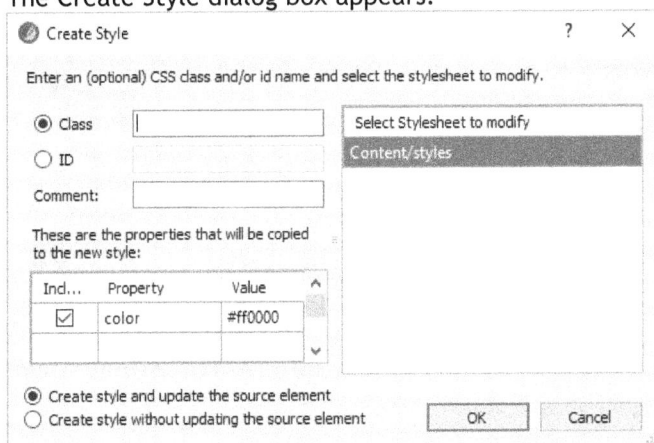

6 In the **Name** field, type a name for the new style without using spaces.

7 If you do not want to include a style property in the new style, deselect its **Include** option.

8 If you want the new style to be applied to the selected content, select **Create style and update the source element**.

If you do not want the new style to be applied to the selected content, select **Create style without updating the source element**.

9 Click **OK**.

The new style is added to the stylesheet.

Removing inline formatting

You can remove inline formatting from a topic to "clean up" the formatting. This feature is especially useful if you import a Word or FrameMaker document that uses a lot of inline formatting.

◇ *Removing inline formatting does not change how your content is tagged: headings, lists, and tables will not become paragraphs.*

To remove inline formatting:

1 Open a topic that contains inline formatting.

2 Highlight the content that contains inline formatting.
You can press **Ctrl+A** to highlight the entire topic.

3 Click **B**.

Converting inline formatting to a style class

You can use the New Styles Suggestions report to convert inline formatting to a style class.

To convert inline formatting to a style class:

1 Select **Analysis** > **Suggestions** > **New Style Suggestions**.
The New Style Suggestions report appears.

2 Highlight the item(s) you want to convert to a style class.

3 Click **Create Style**.
The Create New Style dialog box appears.

4 Type a style **Class Name**.

5 Click **OK**.

Modifying a style

You can use Flare's Stylesheet Editor to modify a style's properties.

To modify a style:

1   Open a stylesheet.
    The Stylesheet Editor appears.

2   Select the style you want to edit.
    If you are using the Advanced view, the selected style's
    properties appear on the right. If you are using the Simplified
    view, double-click the style.

3   Select a property.

4   Select a value to change.

5   Type or select a new value.

6   Modify other properties as needed.

Creating a font set

You can create a font set to specify a list of fonts for a style to use
rather than one font family. If you specify a font that is not installed on
the user's computer, their browser will use a default font (usually
Times). With a font set, you can provide a list of fonts to try before the
default font is used.

If you are developing HTML Help, you should create a font set if you
want to use fonts that are not common in Windows. If you are
developing HTML5 or WebHelp, you should create a font set if you want
to use fonts that are not common in Windows or if your users have Mac
or Linux computers or might be using a mobile device. Fonts that are
commonly installed in Windows might not be available in other
operating systems.

To create a font set:

1   Open a stylesheet.
    The Stylesheet Editor appears.

2 Select **Options** > **Manage Font Sets**.

The Font Set Manager dialog box appears.

3 Click ⊕.

The Define Font Set dialog box appears.

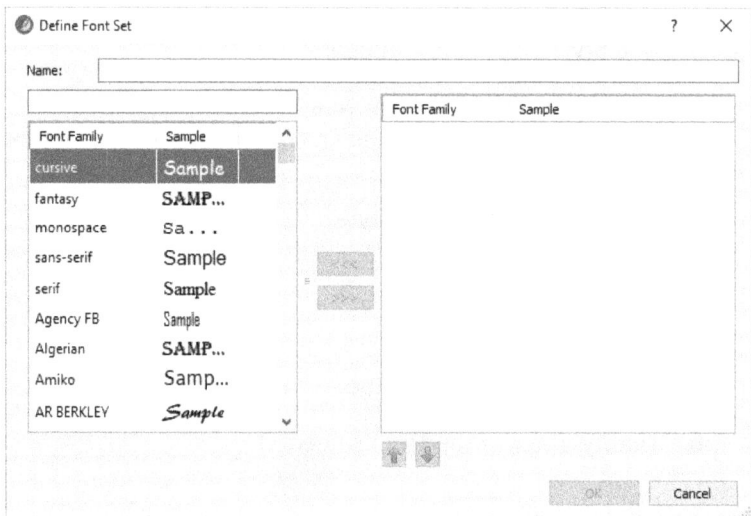

4 Type a **Name** for the font set.

5 In order, select the fonts you want to include and add them to the list.

6 Click **OK** three times.

You can now use your font set when you select a font for your styles.

Creating a CSS variable

You can create a CSS variable to reuse a value in a stylesheet. For example, you can create a CSS variable for a company color and use it for headings, table borders, and other elements. When the color changes, you will only need to update the variable's value rather than updating multiple styles. In Flare 2023 and later, you can also use CSS variables in a table stylesheet or skin.

To create a CSS variable:

1 Open a stylesheet.
   The Stylesheet Editor appears.

2 Select **CSS Variable** > **Add New CSS Variable**.
   The Add New CSS Variable dialog box appears.

3 Select **:root** for the **HTML Element** to create a global variable.
   Global variables can be used for any style in your sylesheet.

4 Type a **Name** for the variable.
   Variable names must begin with - and can only include letters,
   numbers, and underscore.

5 Select a **Property Type**.

6 Type or select a **Value**.

7 Click **OK**.
   The variable is added to your stylesheet.

Using a CSS variable NEW!

If you assigned your CSS variable to the :root element, you can use it
for any style in your stylesheet, in a table stylesheet, or in a skin.

To use a CSS variable:

1 Open a stylesheet in the Advanced View.
  The Stylesheet Editor appears.

2 Select the style you want to edit.

3 Click a style property.

4 Select **CSS Variable** > **Insert CSS Variable**.
  The Insert CSS Variable dialog box appears.

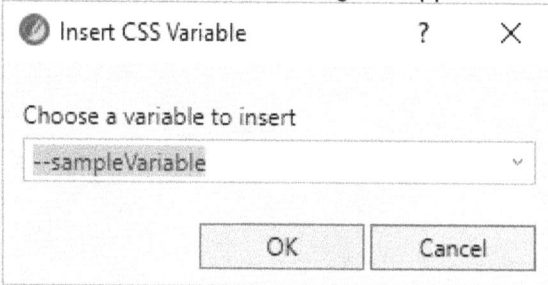

5 Select a variable.

6 Click **OK**.

Modifying a CSS variable

You can modify a CSS variable's name and value. Since CSS variables are considered custom properties, they are modified like other CSS properties.

To modify a CSS variable:

1 Open a stylesheet in the Advanced View.
  The Stylesheet Editor appears.

2 Open the **(Variables)** group at the top of the list of styles on the left side of the Stylesheet Editor.

3 Select the HTML element that is assigned to the variable.
  CSS variables are normally assigned to the :root element.

4 In the list of properties, select the variable.

5 Click 〔···〕.
The Edit CSS Variable dialog box appears.

6 Change the settings as needed.

7 Click **OK**.

Formatting variables

You can use styles to automatically format all variables or a specific variable. Automatically formatting variables using styles is more efficient than applying a style each time you insert a variable.

To format all variables:

1 Open a stylesheet in the Advanced View.
The Stylesheet Editor appears.

2 Select the **MadCap|variable** style.

3 Modify the style's properties.

To format a specific variable:

1 Double-click a variable set.
The Variable Set Editor window appears.

2 Select a variable.

3 Click 〔🗹〕 in the Variable Set Editor toolbar.
The stylesheet opens and a new class is added for the MadCap|variable style.

4 Modify the style's properties.

Using the print medium

Flare's built-in "print" medium is often used when creating print targets. For example, you can set up the print medium to use a different font than online targets.

I often use the print medium to set automatic page breaks for heading 1s and to format my TOC and index styles.

To set automatic page breaks before headings:

1 Open a stylesheet in the Advanced View.
The Stylesheet Editor appears.

2 In the **Medium** drop-down text box, select **Medium: print**.

3 Select a heading style, such as **h1**.

4 Open the **PrintSupport** property group.

5 Change the **page-break-before** option.
You can set the option to always add a page break or to start the heading on the left or right.

To format the TOC and index styles:

1 Open your stylesheet.
Stylesheets are stored in the Resources folder in the Content Explorer.

2 In the **Medium** drop-down text box, select **Medium: print**.

3 Click the plus sign beside the **p** style.

4 Select a TOC or index class.

5 Modify the style's properties. The TOC/index-specific style properties include:

Property	Description
mc-leader-format	Specifies the leader line character. The default is 'dot.' Other options are 'none,' 'dash,' and 'underline.'
mc-leader-indent	Sets the distance between the end of the TOC or index entry and the start of the leader.
mc-leader-offset	Sets the distance between the end of the leader and the page number.

Property	Description
mc-multiline-indent	Specifies additional indentation for TOC or index entries that wrap to more than one line.
mc-pagenum-display	Specifies whether a TOC level should display a page number. If the value is "leaf," the page number will not appear if the TOC entry contains sub-entries.
mc-reference-initial-separator	Specifies the text that appears after index entries and before page numbers. The default is ', '.
mc-reference-separator	Specifies the text that appears between nonconsecutive page numbers. The default is ', '. For consecutive page numbers, Flare will automatically use a dash.

To use the print medium in a target:

1 Open a target.

2 Select the **Advanced** tab.

3 In the **Medium** drop-down text box, select **Medium: print**.

Adding a medium

You can add a medium to a stylesheet to add target-specific formatting. You can use the built-in "tablet" and "mobile" mediums to customize your content's formatting when it is viewed on tablet- and phone-sized screens.

To add a medium:

1 Open your stylesheet.
Stylesheets are stored in the Resources folder in the Content Explorer.

2 In the Stylesheet Editor toolbar, click **Options**.

3 Select **Add Medium**.

4 Type a name for the medium.

5 Click **OK**.

Adding a comment to a style

You can add a comment to a style to document how and/or when it should be used.

To add a comment to a style:

1 Open your stylesheet.

2 If you are using the Simplified View:

- ☐ Select a style.

- ☐ Double-click in the Comment cell.

- ☐ Type the comment.

If you are using the Advanced View:

- ☐ Select a style.

- ☐ Click in the Comment box below the list of styles.

- ☐ Type the comment.

Applying a stylesheet to a topic

Most projects use the same stylesheet for every topic. However, you can create multiple stylesheets and associate different stylesheets with specific topics.

Shortcut	Tool Strip	Ribbon
F4 or Ctrl+Shift+P	🖹 (Content Explorer)	Home > Properties

To assign a stylesheet to a topic:

1 Select a topic and click 🖹.
—OR—
Right-click a topic and select **Properties**.

2 Select the **Topic Properties** tab.

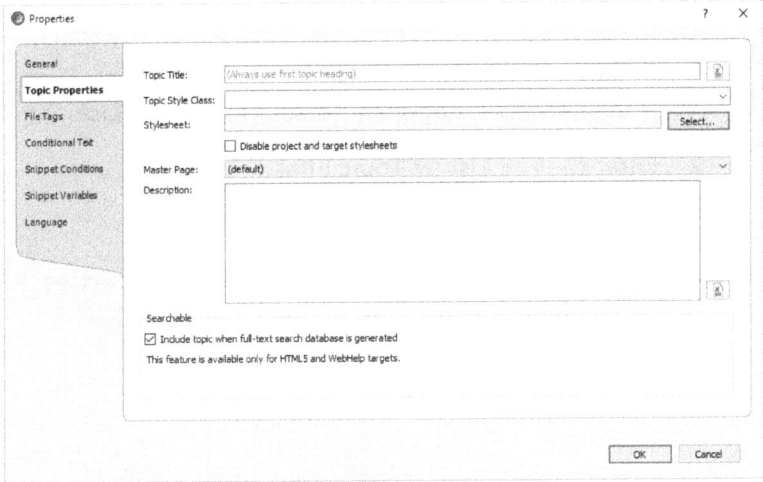

3 Click **Select**.

The Stylesheet Links dialog box appears.

◇ If you cannot select a stylesheet, you will either need to select Disable project and target stylesheets or allow local stylesheets in the Project Properties. See "Applying a stylesheet to all topics" below.

4 Select a **Stylesheet**.

5 Click ⬜.

6 Click **OK**.

The stylesheet is applied to the topic.

Applying a stylesheet to multiple topics

You probably don't want to individually associate each topic with your stylesheet. That could take a *long* time! If you want to use multiple stylesheets in a project, you can associate multiple topics with a stylesheet at the same time using the File List.

◇ If you want to use the same stylesheet for every topic, see "Applying a stylesheet to all topics" on page 257.

To apply a stylesheet to multiple topics:

1 Open the File List.
If the File List is not open, select **View** > **File List**.

2 Filter the File List by **Topic Files**.

3 Select the topics.

4 Right-click the selected topics and select **Properties**.
The Properties dialog box appears.

5 Select the **Topic Properties** tab.

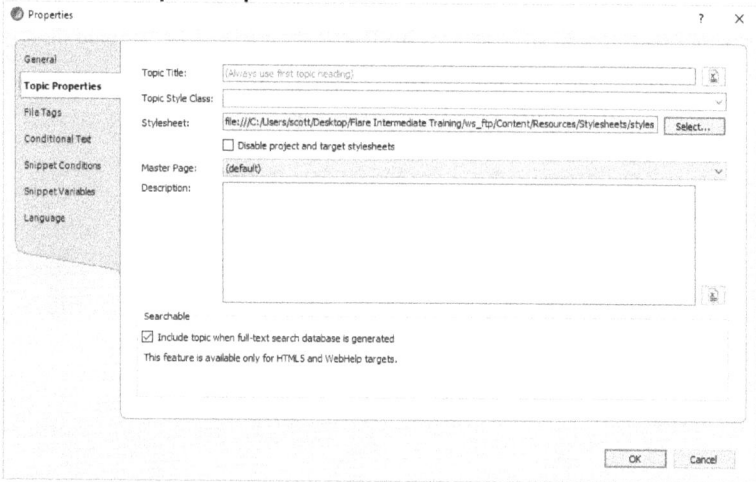

6 Select a **Stylesheet**. If you want to apply multiple stylesheets to the topics, click **Select** and select the stylesheets.

◇ *If you cannot select a stylesheet, you will either need to select Disable project and target stylesheets or allow local stylesheets in the Project Properties. See "Applying a stylesheet to all topics" below.*

7 Click **OK**.
The topics are associated with the stylesheet.

Applying a stylesheet to all topics

If you assign a primary stylesheet to a project, every topic is automatically associated with the primary stylesheet, including imported topics and any new topics you create in the future.

Shortcut	Tool Strip	Ribbon
Alt+P, R	(Project toolbar)	Project > Project Properties

To apply a stylesheet to all topics:

1 Select **Project** > **Project Properties**.
The Project Properties dialog box appears.

2 Select the **Defaults** tab.

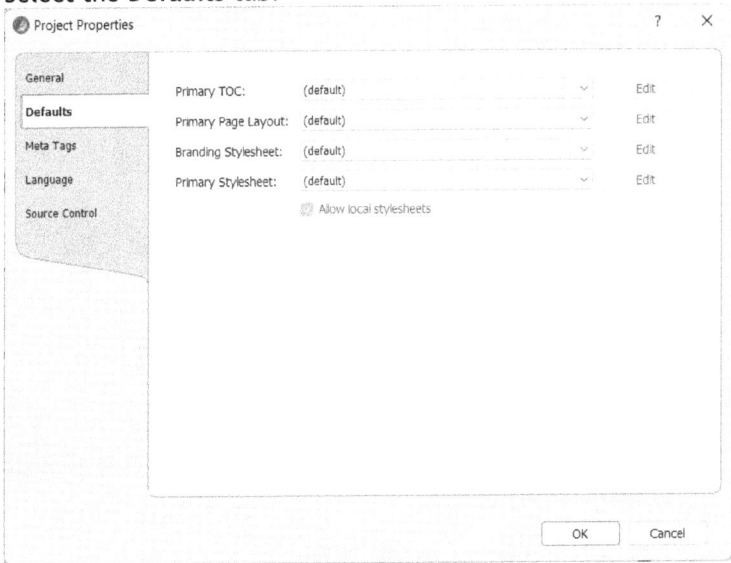

3 Select a **Primary Stylesheet**.

4 If you want to use another stylesheet for some topics in your project, select **Allow local stylesheets**.

5 Click **OK**.

Applying a style to content

You can apply a style to any type of content, including text and images, and dynamic content such as drop-down and toggle links.

◇ *If you want to apply a style to a table, you should use a table style. See "Applying a table style to a table" on page 125.*

To apply a style:

1 Select the content you want to format. If you want to format multiple blocks of content, such as paragraphs, highlight all of the paragraphs.

2 Select **Home** > **Style Window**.
   —OR—
   Select **Home** and open the Style drop-down text box.
   —OR—
   Press **F12**.

3 Click a style's name in the list.

Reviewing formatting applied to content

You can use the Formatting window to review and change formatting that has been applied by styles and inline formatting.

To review formatting applied to content:

1 Click inside a block of content or highlight content such as an image or word.

2 Select **Home** > **Formatting Window**.
   —OR—
   Press **Ctrl+F12**.
   The Formatting window appears.

3 Select the **Style Inspector** tab to review formatting applied from style settings.
   —OR—
   Select the **Local Style Properties** tab to review formatting applied from inline formatting.

4 To change the formatting, select a style setting and click ⋯ .

5 To add formatting, click **+**.

Pinning a style

You can pin frequently used styles to the top of the style list. Pinning a style makes it easier to find.

To pin a style in the style list:

1 Select **Home** > **Style Window**.
—OR—
Select **Home** and open the Style drop-down text box.
—OR—
Press **F12**.

2 Hover over a style in the list.

3 Click ▪ .
The style is pinned.

Branding stylesheets NEW!

You can use a branding stylesheet to store your company's colors, preferred font, and logo as CSS variables. You can then use these branding CSS variables in your stylesheets, table stylesheets, and skins.

For more information about creating and using CSS variables, see "Using a CSS Variable" on page 249.

Creating a branding stylesheet

You can create multiple branding stylesheets if you need to use different branding elements (such as colors) for different targets.

Shortcut	Tool Strip	Ribbon
Ctrl+T	▤ (Content Explorer)	File > New

To create a branding stylesheet:

1 Select **File** > **New**.
   The Add File dialog box appears.

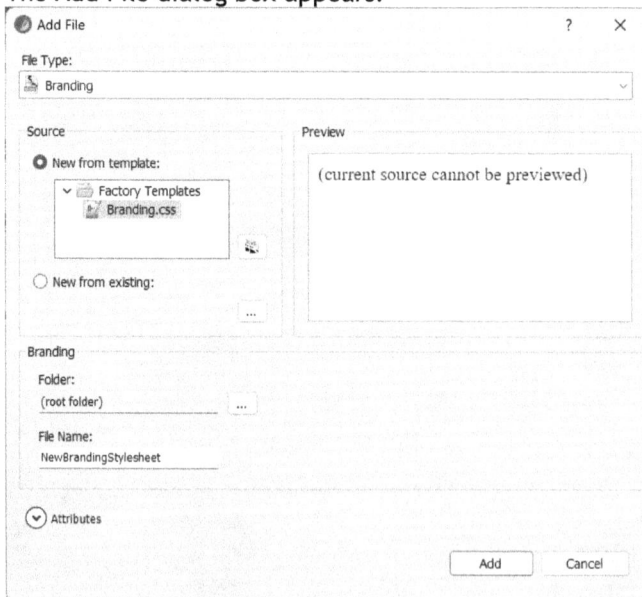

2 For **File Type**, select Branding.

3 Select a **Source** template.

4 Type a **File Name** for the branding stylesheet.
Branding stylesheets have a .css extension. If you don't type the extension, Flare will add it for you.

5 Click **Add**.
The branding stylesheet appears in the Content Explorer and opens in the Branding Editor.

Modifying a branding stylesheet

To modify a branding stylesheet:

1 Open a branding stylesheet.
The Branding Editor appears.

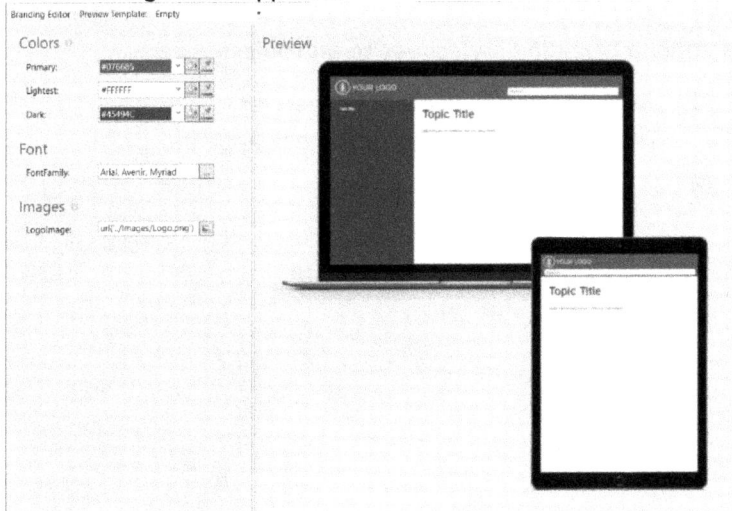

2 In the **Colors** section, select a value for the following CSS variables:

☐ Primary

☐ Lightest

☐ Dark

3 In the **Font** section, select a font or font set for the **FontFamily** CSS variable.

4 In the **Images** section, select an image to use for the **LogoImage** CSS variable.

Applying a branding stylesheet to all topics

If you assign a branding stylesheet to a project, every topic is automatically associated with the branding stylesheet, including imported topics and any new topics you create in the future. The branding stylesheet will also be used for every target unless you select a different branding stylesheet in the target.

Shortcut	Tool Strip	Ribbon
Alt+P, R	🖳 (Project toolbar)	Project > Project Properties

To apply a branding stylesheet to all topics:

1 Select **Project** > **Project Properties**.
The Project Properties dialog box appears.

2 Select the **Defaults** tab.

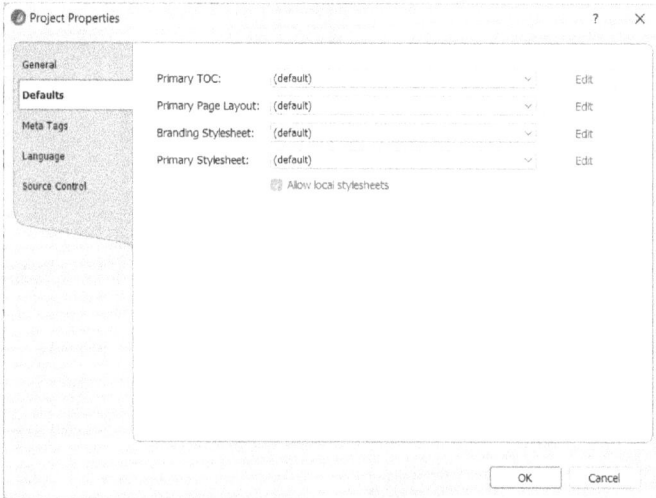

3 Select a **Branding Stylesheet**.

4 Click **OK**.

Selecting a branding stylesheet in a target

If you select a branding stylesheet in a target, the target will use the target's branding stylesheet instead of the branding stylesheet selected for the project.

Shortcut	Tool Strip	Ribbon
Alt+P, R	(Project toolbar)	Project > Project Properties

To select a branding stylesheet in a target:

1 Open a target.

2 Select the **General** tab.

3 Select a **Branding Stylesheet**.

Page layouts

You can create a page layout to set the page size and margins and to add headers and footers to print targets. You can add pages to a page layout to set up different headers and footers for first, title, left, and odd pages.

Creating a page layout

You can create multiple page layouts to apply different page sizes, margins, headers, or footers to different sections in a print target. For example, you can create a landscape page layout for wide tables or large graphics.

Page layouts have a .flpgl extension, and they are stored in the Resources\PageLayouts folder in the Content Explorer.

To create a page layout:

1 Select **File** > **New**.
The Add File dialog box appears.

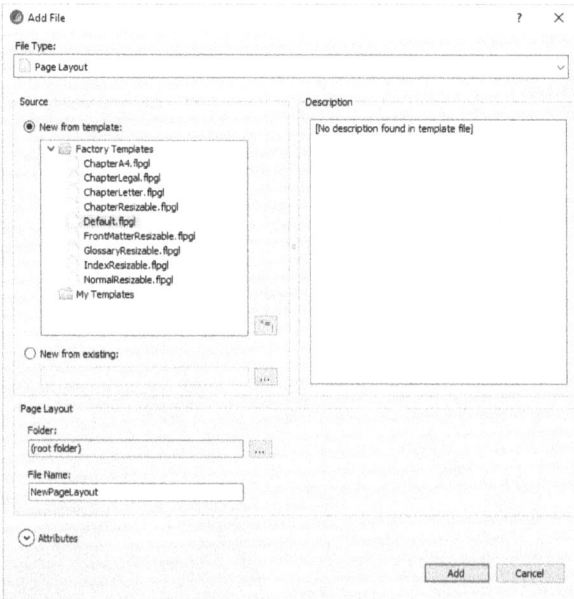

2 For **File Type**, select **Page Layout**.

3 Select a **Source** template.

4 Type a **File Name**.

5 Click **Add**.
  The page layout appears in the Page Layout Editor.

Adding pages to a page layout

You can include the following types of pages in a page layout:

Page Type	Description
Title	The title page is often used for a cover or title page, which is also usually the first page in your TOC.
First First Left First Right	The first page is often used for the first page in a chapter. For more information about adding chapter breaks, see "Specifying chapter breaks" on page 194.
Left	The left page is automatically used for the left (even) pages.
Right	The right page is automatically used for the right (odd) pages.
Empty Empty Left Empty Right	Empty pages are inserted as needed if you set up chapters to end on left pages or set up styles (such as headings) to start on left or right pages. If you do not have an empty page in your page layout, the inserted page will be blank.

TIP *If you are creating left and right pages for a Word target, you should also select the "Generate 'Mirror Margins' for MS Word Output" option on the Advanced tab in your Word target.*

To add a page to a page layout:

1 Open a page layout.

2 In the Page Layout Editor toolbar, click and select **Add Page**.
  The new page appears.

3 Right-click the page and select **Page Properties**.

The Page Properties dialog box appears.

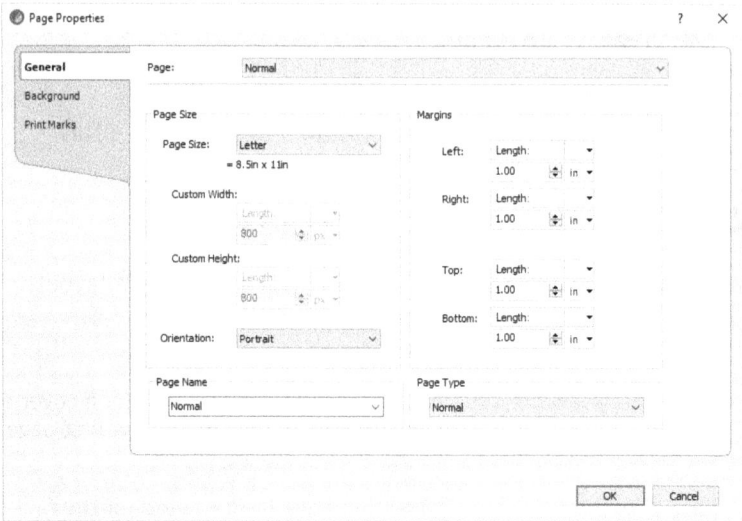

4 Select a **Page Type**.

5 Type a **Page Name**.

6 Click **OK**.

Duplicating a page

Instead of adding pages, you can duplicate a page in a page layout. If you add a page, you will need to add and set up the frames in the page. Duplicate a page is usually easier, since the frames are already set up.

To duplicate a page:

1 Open a page layout.

2 Right-click a page icon and select **Duplicate Page**.

3 Set up the new page.

Adding a frame to a page layout

You can add a decoration or image frame to include text or images on a page. For example, you can add a decoration frame to include the word "draft" on a page.

To add a decoration frame to a page layout:

1 Open a page layout.

2 Select a page.

3 Click .

4 Select **New Decoration Frame Mode**.
 The cursor changes to a crosshair.

5 Click and drag to draw the decoration frame.

6 Right-click inside the decoration frame and select **Edit Text**.

7 If you want to use a template, select **Yes** and select a template.

8 Add content to the frame.
 You can include any type of content in the frame, including text, images, tables, snippets, and variables.

To add an image frame to a page layout:

1 Open a page layout.

2 Select a page.

3 Click .

4 Select **New Image Frame**.
 The Insert Image dialog box appears.

5 Select an image.

6 Position the image frame on the page.

Adding running headings to a header or footer

You can use Flare's built-in heading variables to add running headings to a header or footer. A "running" header variable automatically updates whenever the specified heading level appears. For example, the footer in each chapter of this book automatically changes for each heading level 1.

To add a running heading to a header or footer:

1 Open a page layout.

2 Right-click a header or footer and select **Edit Text**. The header or footer pane appears.

3 Select **Insert > Variable**.

4 In the **Variable Sets** column, select **Heading**.

5 Select one of the "Level" variables. The level1-6 variables match the h1-h6 styles.

6 Click **OK**.

Adding page numbers to a header or footer

You can use Flare's built-in system variables to add page numbers to a header or footer.

To add page numbers to a header or footer:

1 Open a page layout.

2 Right-click a header or footer and select **Edit Text**. The header or footer pane appears.

3 Select **Insert > Variable**.

4 In the **Variable Sets** column, select **System**.

5 Select the **PageNumber** variable.

*You can use the **PageCount** variable to insert the total page count.*

6   Click **OK**.

Creating a two-column page layout

You can add another body frame to a page layout to create a two-column page layout.

To create a two-column page layout:

1   Open a page layout.

2   Select a page.

3   Select the **body** frame.

4   Size the body frame to make space for the second column.

5   Right-click the body frame and select **Copy**.

6   Right-click the body frame again and select **Paste**.

7   Position the new body frame. You will see an arrow between the two body frames that indicates how the content will flow between the frames.

Rotating frames in a page layout

You can rotate a frame clockwise or counterclockwise in 45-degree increments.

To rotate a frame:

1   Open a page layout.

2   Select a page.

3   Select a frame.

4   Click **Layout**.

5   Select **Rotate** and select a **Rotation** option.

Stacking frames in a page layout

If your frames overlap, you can set their stacking order by moving them above or below each other.

To move a frame below or above another frame:

1 Open a page layout.

2 Select a page.

3 Select a frame.

4 Click **Layout**.

5 Select **Depth** and select whether the frame should move above (float) or below (sink) other overlapping frames.

Applying a page layout to a topic

To assign a page type to a topic, you need to add a chapter or page layout break in your TOC.

To add a break in a TOC:

1 Open a TOC.

2 Right-click a TOC book or page and select **Properties**.
The Properties dialog box appears.

3 Select the **Printed Output** tab.

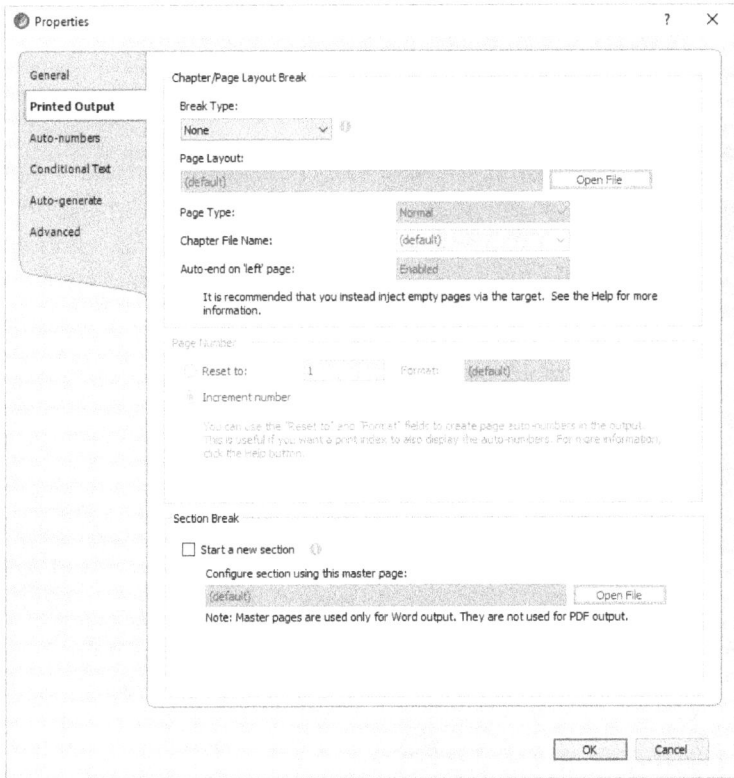

4 For **Break Type**, select **Chapter Break or Page Layout Break**. If you want to restart the page numbering or change the page number format, select **Chapter Break**. If not, select **Page Layout Break**.

5 Select a **Page Layout**.

6 Select a **Page Type**.

7 Select whether you want the chapter to **Auto-end on 'left' page**.

8 Click **OK**.

Applying a page layout to a target

1 Open a target.

2 On the **General** tab, select a **Primary Page Layout**.

Template pages

You can create a template page to include content at the top or bottom of topics in online targets, such as WebHelp. In addition to text, images, tables and other content, you can include the following dynamic elements (Flare calls these elements "proxies") in a template page:

- **Breadcrumbs** automatically adds the path to the current topic using the "books" in your TOC/menu. If your books are linked to topics, the books in the breadcrumb path are also linked.

- A **mini-TOC** automatically adds a list of links to "book-level" topics in your TOC. For example, if a TOC book links to the "Overview" topic and the book contains pages that link to topics A, B, and C, a mini-TOC would add links to topics A, B, and C to the Overview topic.

Creating a template page

You can create as many template pages as you need. Template pages have a .flmsp extension, and they are stored in the Resources\TemplatePages folder in the Content Explorer.

Shortcut	Tool Strip	Ribbon
Ctrl+T	📑 (Content Explorer	File > New

To create a template page:

1   Select **File** > **New**.
    The Add File dialog box appears.

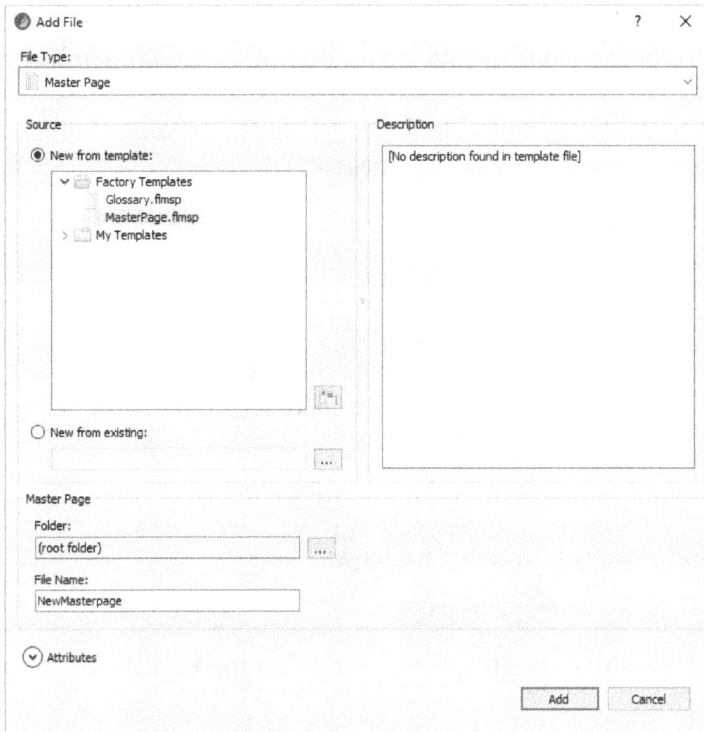

2   For **File Type**, select **Template Page**.

3   Select a **Source** template.

4   Type a **File name**.
    Template pages have a .flmsp extension. If you don't type the
    extension, Flare will add it for you.

5   Click **Add**.
    The template page appears in the Content Explorer and opens
    in the XML Editor.

Adding content to a template page

You can add content to a template page above or below the topic body
proxy.

Any content that you add above the topic body proxy will appear above
the topic's content. However, it is not "fixed" on the screen: it will
scroll off the page in long topics.

Any content that you add below the topic body proxy will appear at the bottom of the topic. In long topics, users may need to scroll down to see it.

To add content to a template page:

1 Open a template page.

2 Position the cursor above or below the **topic body** proxy.

3 Type or insert your content.
You can add anything to a template page that you can add to a topic, including formatted content, images, lists, tables, variables, and snippets.

To add a proxy to a template page:

1 Open a template page.

2 Position the cursor above or below the **topic body** proxy.

3 Select **Insert** > **Proxy** and select a proxy.

Applying a template page to a target

To use a template page, you need to associate it with a target. You can associate the same template page with multiple targets, or you can associate a different template page with each of your targets.

To apply a template page to a target:

1 Open a target.

2 Select the **Advanced** tab.

3 Select a **Template Page**.

4 Click **Save**.

Applying a template page to a topic

If you want to use a different template page for a topic or a group of topics, you can apply a template page at the topic level. For example, many Flare users use a different template page for their "Home" page.

To apply a template page to a topic:

1 Right-click a topic in the Content Explorer or File List and select **Properties**.
   —OR—
   Select a topic and press **F4**.

2 Select the **Topic Properties** tab.

3 Select a **Template Page**.

4 Click **OK**.

Skins

Skins are used to format the banner (or "header"), toolbar, and menu in online targets. The HTML5 format supports the most customization options.

Skin files have a .flskn extension. They appear in the Skins folder in the Project Organizer.

'Is there a skin gallery?'

You can download and customize skins from MadCap's website at www.madcapsoftware.com/downloads/flare-html5-skins.aspx.

Creating a skin

You can use the same skin for all of your online targets, or you can create different skins for each target.

Shortcut	Tool Strip	Ribbon
Ctrl+T	📑 (Content Explorer	File > New

To create a skin:

1   Select **File** > **New**.
    —OR—
    Right-click the **Skins** folder and select **Add Skin**.
    The Add File dialog box appears.

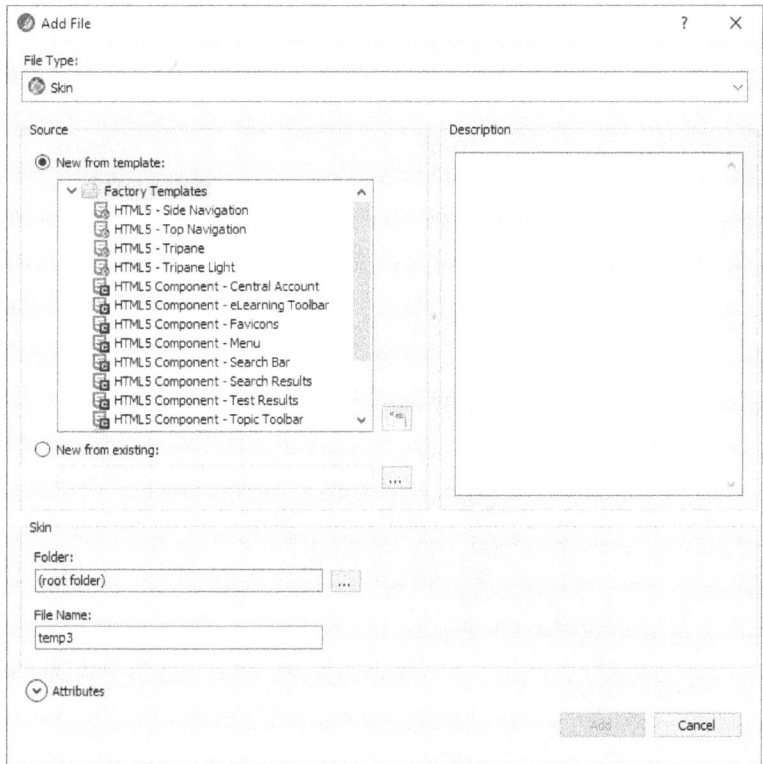

2 For **File Type**, select **Skin**.

3 Select a **Source** template.

4 Type a **File Name**.

Skins have a .flskn extension. If you don't type the extension, Flare will add it for you.

5 Click **Add**.

The skin appears in the Skins folder in the Project Organizer and opens in the Skin Editor.

Modifying an HTML5 skin

You can modify a skin to change the size, appearance, and features used in your HTML5 targets. HTML5 has four skin design options: tripane, top navigation ("topnav"), side navigation ("sidenav"), and skin components.

The original skin design is named "tripane" based on its toolbar, navigation, and content panes. The top and side navigation designs provide a content pane and a navigation pane that typically includes a logo, menu (instead of a TOC), and search bar. The sidenav option provides a navigation "tree" menu pane like the HTML5 tripane and WebHelp designs. The fourth option, skin components, can be used to create a very customized design. Rather than starting with a pre-set layout, you can use proxies to create your own layout and skin components to customize the proxies.

To modify an HTML5 topnav or sidenav skin:

1　Open a skin.
　　The skin appears in the HTML5 Skin Editor.

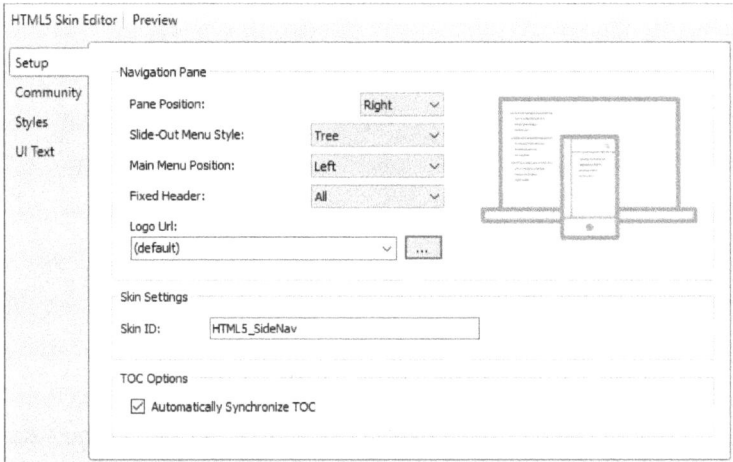

2　Select the **Setup** tab.

　　□　Select a **Pane Position.**

　　□　Select a **Slide-out Menu Style.**
　　　　The slide-out menu is used to display the menu on smaller screens. "Tree" displays the full menu and the sub-menus can be expanded or collapsed. "Drilldown" only displays the current sub-menu.

　　□　Select a **Main Menu Position.**

　　□　If you set the **Main Menu Position** to "top," select a **Top Menu Levels to Show** depth.

□ Select if you want to use a **Fixed Header** for web (desktop), tablet, and/or mobile layouts. A fixed header will not scroll.

□ Select a **Logo URL.**

□ Enable or disable **Automatically Synchronize TOC** for the side and slide-out menus.

3   Select the **Styles** tab and modify the styles as needed. You can change fonts, colors, icons, labels, borders, paddings, and background settings and other style properties for the skin elements.

TIP▶ *You can use the filter textbox to quickly find a style or property. Or, you can click **Highlight** in the Skin Editor toolbar and click the element you want to change in the preview window.*

4   Select the **UI Text** tab and modify any of the labels or tooltips.

5   Click **Save.**

To modify an HTML5 tripane skin:

1   Open a skin.
    The skin appears in the HTML5 Skin Editor.

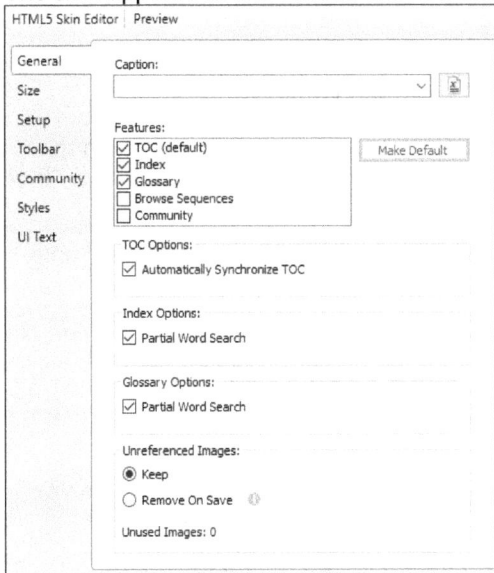

2 On the **General** tab, type a **Caption**.

The caption appears in the browser's title bar or tab.

3 Select the **Features** that you want to appear in the help window.

4 Enable or disable **Automatically Synchronize TOC**.

5 Select the **Size** tab.

6 If you want to specify a size for your help window:

□ Deselect **Use Browser Default Size**.

□ Type values for the window positions.
 —OR—
 Click **Preview Full Size**, resize the preview window, and click **OK**.

7 Select the **Setup** tab.

□ Select a **Pane Position**.

□ Type a **Pane Size**.

□ If you want your content to resize and adapt to different screen sizes, select **Enable responsive output**.

8 Select the **Toolbar** tab and select the **Toolbar Buttons** you want to include:

Button	Description
Current Topic Index	The current topic's position in the browse sequence. For example, "Page 2 of 5."
Edit User Profile	Opens the Feedback Service Profile dialog box.
Expand All	Expands (or collapses) all toggler, drop-down, and expanding links in the current topic.
Filler	Adds space between buttons
Next Topic	Opens the next topic in the browse sequence.

Button	Description
Previous Topic	Opens the previous topic in the browse sequence.
Print	Opens the Print dialog box.
Remove Highlight	Hides the search term highlighting in the topic.
Select Language	Allows the user to switch to another language if you build multilingual HTML5 targets that all contain the Select Language button.
Select Skin	Allows the user to switch to another language. See "Enabling runtime skins" on page 286.
Separator	Adds a dividing line between buttons.
Topic Ratings	Shows the MadCap Pulse topic rating icons (stars by default).

9 Select the **Styles** tab and modify the styles as needed. You can change fonts, icons, labels, borders, paddings, and background settings and other style properties for the skin elements.

TIP▶ *To change or remove the MadCap logo, open the* **Header** *style group, select* **Logo**, *select* **Background**, *and change the* **Image** *setting.*

10 Select the **UI Text** tab and modify any of the labels or tooltips.

11 Click **Save**.

Modifying a WebHelp or HTML Help skin

You can modify a skin to change the size, appearance, and features used in your WebHelp or HTML Help targets.

To modify a skin:

1 Open a skin.

The skin appears in the Skin Editor.

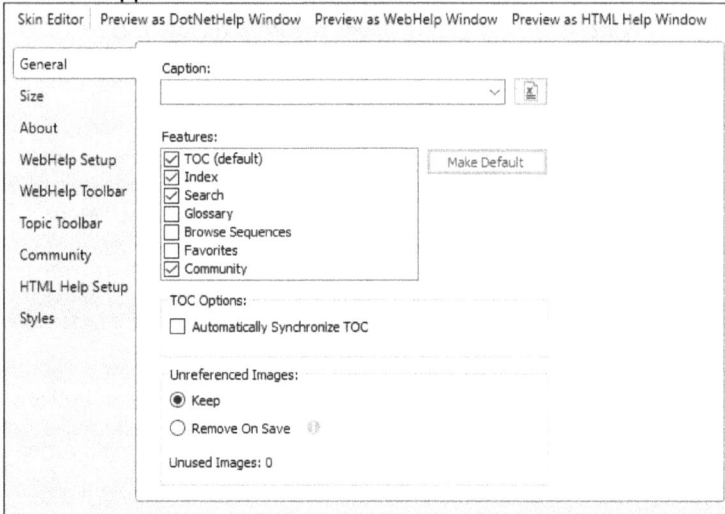

| Skin Editor | Preview as DotNetHelp Window | Preview as WebHelp Window | Preview as HTML Help Window |

General
Size
About
WebHelp Setup
WebHelp Toolbar
Topic Toolbar
Community
HTML Help Setup
Styles

Caption:

[_____] ⌄ 🗵

Features:

☑ TOC (default)   Make Default
☑ Index
☑ Search
☐ Glossary
☐ Browse Sequences
☐ Favorites
☑ Community

TOC Options:
☐ Automatically Synchronize TOC

Unreferenced Images:
◉ Keep
○ Remove On Save

Unused Images: 0

2 On the **General** tab, type a **Caption**.

The caption appears in the browser window's title bar or tab.

3 Select the **Features** that you want to appear in the help window.

4 Enable or disable **Automatically Synchronize TOC**.

5 Select the **Size** tab.

6 If you want to specify a size for your help window:

☐ Deselect **Use Browser Default Size**.

☐ Type values for the window positions.
 —OR—
 Click **Preview Full Size**, resize the preview window, and click **OK**.

7 If you are editing a WebHelp skin:

☐ Select the **About** tab.

- [ ] Select an **About** box image.

  The About box appears when the user clicks the logo on the far right of the toolbar.

- [ ] Select the **WebHelp Setup** tab.

- [ ] Select a **Pane Position**.

- [ ] Type a **Navigation Pane Size**.

- [ ] Select the number of **Visible Accordion Items** you would like to use (navigational features appear in an "accordion," similar to tabs). If you have selected more navigational features on the **General** tab than visible accordion items, the additional items will appear as icons below the accordion.

- [ ] Select the **WebHelp Toolbar** tab.

- [ ] Select the **Toolbar Buttons** you want to include:

Button	Description
Add Topic to Favorites	Adds the current topic to the Favorites list.
Back	Opens the previously viewed topic.
Collapse All	Collapses all toggler, drop-down, and expanding links in the current topic.
Current Topic Index	The current topic's position in the browse sequence. For example, "Page 2 of 5."
Edit User Profile	Opens the Feedback Service Profile dialog box.
Expand All	Expands all toggler, drop-down, and expanding links in the current topic.
Forward	Opens the next topic (if the user has previously clicked the Back button).
Home	Opens the startup topic as specified on the Target's General tab.
Next Topic	Opens the next topic in the browse sequence.
Previous Topic	Opens the previous topic in the browse sequence.

Button	Description
Print	Open the Print dialog box.
Quick Search	Searches the current topic for a word or phrase.
Refresh	Reopens the current topic.
Remove Highlight	Turns off search highlighting.
Select Browse Sequence	Open the browse sequence in the navigation pane.
Select Favorites	Open the favorites list in the navigation pane.
Select Glossary	Opens the glossary in the navigation pane.
Select Index	Opens the index in the navigation pane.
Select Search	Opens the search in the navigation pane.
Select Skin	Allows the user to switch to another language. See "Enabling runtime skins" on page 286.
Select TOC	Opens table of contents in the navigation pane.
Separator	Adds a dividing line between buttons.
Stop	Cancels opening the topic.
Toggle Navigation Pane	Hides and shows the navigation pane.
Topic Ratings	Shows the MadCap Pulse topic rating icons (stars by default).

8 If you are editing an HTML Help skin:

☐ Click the **HTML Help Setup** tab.

☐ Select the **HTML Help Buttons** you want to include:

Button	Description
Hide	Hides and shows the navigation pane.
Locate	Highlights the current topic in the TOC.
Back	Opens the previously viewed topic.

Button	Description
Forward	Opens the next topic (if the user has previously clicked the Back button).
Stop	Cancels opening the topic.
Refresh	Reopens the current topic.
Home	Opens the startup topic as specified on the Target's General tab.
Font	Increases the font size in topics.
Print	Open the Print dialog box.
QuickSearch	Searches the current topic.
Next	Opens the next topic in the TOC.
Previous	Opens the previous topic in the TOC.
Options	Opens a menu with the following commands: Home, Show, Back, Stop, Refresh, and Search Highlight On/Off.
Jump 1 and 2	Opens a specified website or topic (click Jump Button Options to select a target).

☐ Select the **Button**, **Navigation Pane**, and **Misc Options** you want to use.

TIP▶ *To include the WebHelp toolbar (created on the WebHelp Toolbar tab) in HTML Help, select **Display Toolbar in Each Topic**.*

9 If you are editing a WebHelp skin, select the **Styles** tab and modify the styles as needed. You can change the font, icon, label, border, padding, and background settings for any toolbar item.

TIP▶ *To change or remove the MadCap logo, open the **ToolbarItem** style group, select **Logo**, and change the **Icon** setting.*

10 Click **Save**.

Applying a skin to a target

To use a skin, you need to apply it to a target. You can apply the same skin to multipleeLearning targets, or you can apply different skins to each target.

To apply a skin to a target:

1 Open a target.

2 If you want to apply a skin to an HTML5 target:

- □ Select the **Skin** tab.

- □ Select a **Skin**.

- □ If you are using skin components, select your skin components.

If you want to apply a skin to any other type of online target:

- □ Select the **General** tab.

- □ Select a **Skin**.

3 Click **Save**.

Enabling runtime skins

You can enable runtime skins for HTML5 targets to allow the user to select a skin. For example, you could allow the user to select a skin with or without a toolbar or a "dark" or "light" design.

◇ *You can allow the user to select different topnav or tripane skins, but runtime skins cannot switch from a topnav to tripane skin.*

To enable runtime skins:

1 Create the skins you want to use.

2 If you are setting up runtime skins for a tripane HTML5 target:

- □ Open each tripane skin.

- □ Select the **Toolbar** tab.

□ Add the **Select Skin** toolbar button.

If you are setting up runtime skins for a topnav HTML5 target:

□ Open or create a topic toolbar skin.

□ Select the **Setup** tab.

□ Add the **Select Skin** toolbar button.

3 Open an HTML5 target.

4 Select the **Advanced** tab.

5 Select **Generate All Skins**.

Responsive design

Flare uses a responsive layout framework based on a simplified version of ZURB's Foundation framework. Like Foundation, the Flare framework divides the screen into a 12-column grid. These columns are used to specify widths. For example, 12 columns equal 100%, 6 columns equal 50%, and 4 columns equal 33.33%.

The column widths automatically adjust based on the width of the screen. When you set up a responsive layout, you can specify different column widths for desktops, tablets, and phones. If the screen is narrow (e.g., on a phone), you can set the columns in a row to automatically stack top to bottom rather than left to right.

Creating a responsive layout

You can create as many responsive layouts as you need. For example, you could create a three-column responsive layout to display links on a home page and a two-column layout to display side-by-side screenshots and procedures in topics.

Shortcut	Tool Strip	Ribbon
Alt+H, R	none	Home > Responsive Layout

To create a responsive layout:

1   Open a template page, topic, or snippet.

2   Position the cursor where you want to insert the responsive layout.

3 Select **Home** > **Responsive Layout.**

The Responsive Layout pane appears.

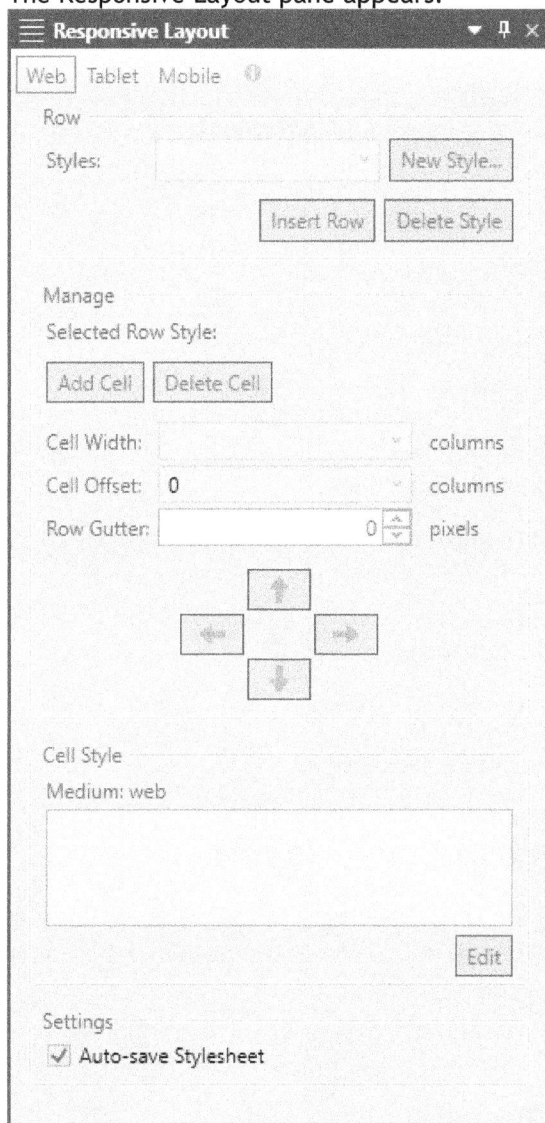

4 Click **New Style.**

5 Type a **Class Name.**

6 Select a **Stylesheet.**

7 Select a **Row Template.**

8 Click **OK.**

Flare creates the responsive layout style definitions in the selected stylesheet.

9 Click **Insert Row.**

The responsive layout is added to your template page, topic, or snippet.

10 Click inside the first cell.

11 Click **Web.**

12 Select a **Cell Width.**

13 Select a **Cell Offset.**

14 In you have additional cells, click inside each cell and select a **Cell Width** and a **Cell Offset.**

15 If you need to add a row, click **Insert Row** and set up the cells.

16 Click **Tablet.**

17 Set the cell widths and offsets.

18 Click **Mobile.**

19 Set the cell widths and offsets.

Inserting a responsive layout

Responsive layouts are often inserted into template pages so that they can be used in multiple topics. However, you can insert a template page into a topic or snippet.

Shortcut	Tool Strip	Ribbon
Alt+H, R	none	Home > Responsive Layout

To insert a responsive layout:

1 Open a template page, topic, or snippet.

2 Position the cursor where you want to insert the responsive layout.

3 Select **Home** > **Responsive Layout**.
The Responsive Layout pane appears.

4 Select a **Style**.

5 Click **Insert Row**.
The responsive layout is added to your template page, topic, or snippet.

Enabling responsive output

Responsive output is automatically enabled in HTML5 topnav and sidenav skins. However, you must enable it in HTML5 tripane skins.

To enable responsive output:

1 Open an HTML5 tripane skin.
The HTML5 Skin Editor appears.

2 Select the **Setup** tab.

3 Select **Enable Responsive Output**.

4 If you want to change the maximum width settings for tablets or mobile devices, type a size.

Designing responsive output

You can use the Web, Tablet, and Mobile mediums in an HTML5 skin to specify how your content will appear on each type of device.

To design responsive output:

1 Open an HTML5 skin.
The HTML5 Skin Editor appears.

2 Select the **Styles** tab.

3 In the HTML5 Skin Editor toolbar, click **Web Medium**.

4 Set the colors, icons, fonts, and other settings for the web medium. The web medium will be used when your content is viewed on a PC.

5   Click **Table Medium**.

The Tablet medium inherits all of the settings from the Web medium. The only elements that are automatically different are the navigation icons since they are different sizes.

6   Set any design elements that should be different in the tablet medium.

7   Click **Mobile Medium**.

The Mobile medium inherits all of the settings from the Tablet medium. The only elements that are automatically different are the navigation icons since they are different sizes.

8   Set any design elements that should be different in the mobile medium.

Viewing responsive output in the XML Editor

You can use the layout resizer to test how a topic will appear at different screen widths.

To view responsive output using the layout resizer:

1   Build an HTML5 target.

2   Click ⊨⊣ in the XML Editor's lower toolbar. The layout resizer bar appears at the top of the editor window.

3   Click a location in the slider or drag the preset markers to see how the topic will appear with different browser widths.

Viewing responsive output

After you build an HTML5 target, you can view it as it will appear on a tablet or phone.

To view responsive output:

1   Build an HTML5 target.

2   In the Build Progress dialog box, click **View Output**.

3 Select **Smartphone** or **Tablet**.
   The content appears in a preview window that looks like the selected device.

When you view the HTML5 target using the Smartphone or Tablet option, it does not open in an emulator. Instead, it opens in your browser inside a picture of a tablet or phone. You can use these views to review the design, but they are not for testing. To test your content, you should open it using a tablet and/or phone or use an emulator in your browser.

Using responsive condition tags

The default "MyConditionTags" condition tag template includes three responsive condition tags: Web, Tablet, Mobile. You can use these condition tags in HTML5 targets to show or hide content based on the screen width. For example, a topic might contain a screenshot for small screens and a video for larger screens. You could apply the "Mobile" condition tag to the screenshot and the "Tablet" and "Web" condition tags to the video.

Responsive conditions can be applied to content inside a topic, snippet, template page, or micro content response. They should not be applied to files or folders.

For more information about condition tags, see "Condition tags" on page 313.

To apply a responsive condition tag to content:

1 Open a topic.

2 Select the content to be tagged.
   You can apply a tag to any content, including characters, words, paragraphs, tables, items in a list, and images.
   ◇ *You can apply a responsive condition tag to an entire table, but you cannot apply a responsive condition to a table row, column, or cell if the table uses a table style.*

3 Select **Home** > **Conditions**.
   —OR—

Press **Ctrl+Shift+C**.

The Condition Tags dialog box appears.

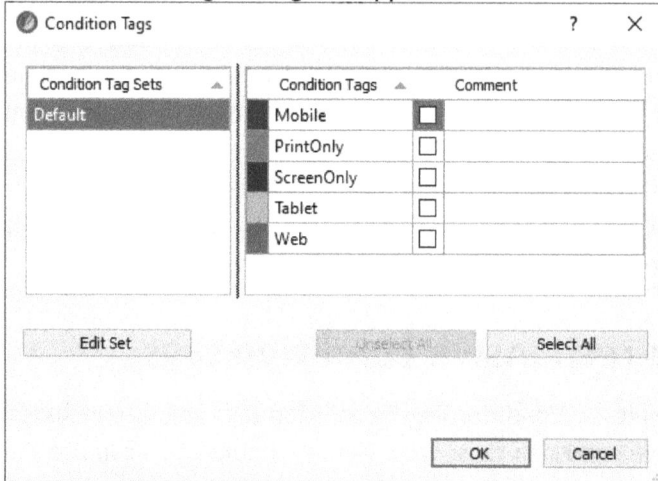

4 Select the Web, Tablet, and/or Mobile checkbox.

5 Click **OK**.

The condition tag is applied. The tagged content is shaded using the tag's color.

Sample questions for this section

1 What is inline formatting?
A) Formatting that is applied to a word or phrase rather than an entire block of content.
B) Formatting that is applied by highlighting content and manually changing its appearance.
C) "Track changes" lines that automatically appear for new or modified content.
D) Formatting that applies a strikethrough line to your content.

2 You should consider using a font set if you plan to create which format?
A) Eclipse Help
B) Microsoft Word
C) HTML5
D) PDF

3 If you select a primary stylesheet, it is automatically assigned to which topics?
A) Existing topics
B) Imported topics
C) New topics you create after selecting the primary stylesheet.
D) All of the above

4 How can you set different odd and even footers for a print target?
A) Create two template pages.
B) Create two page layouts.
C) Create a page layout with two template pages.
D) Create an odd and even page in a page layout.

5 What can you do with a page layout?
A) Add content to all topics, such as a breadcrumb path.
B) Specify different topic layouts for desktops, tablets, and phones.
C) Specify the page size and margins for a print target.
D) Create templates for new topics that contain sample text and images.

6 What is a responsive layout?
A) A design that automatically changes based on the user's job

function.

B) A layout that can adjust its width based on the screen width.

C) A form designer you can use to add a feedback form to topics.

D) A discussion group you can add to the bottom of topics in online targets.

7 What is a breadcrumb?

A) The path to the current topic using the TOC.

B) The list of previous topics the user has viewed.

C) A way to mark a topic, like a favorite, so that it can be quickly found again.

D) A user comment about a topic.

8 How do you apply a skin to an HTML5 target?

A) Open the skin and select the target on the Targets tab.

B) Open the target and select the skin on the Skins tab.

C) Select File > Options and select a Skin.

D) Select Project Properties and select the skin as the Primary Skin.

Single source

This section covers:

- Variables
- Snippets
- Condition tags
- Micro content

Variables

A variable can only contain unformatted text. I often use variables for copyright statements and product names. If the copyright statement or product name changes, I just change my variable's definition and all of my topics are updated.

Flare also provides dynamic system and heading variables. System variables can be used to insert the date, time, page count, page number, and topic title. Heading variables can be used to insert the current heading (any level or a specific level). System and heading variables are often used in page layouts to set up headers and footers for print targets.

User-defined variables are stored in a variable set. You can create as many variable sets and variables as you need. Variable sets are stored in a file with a .flvar extension. They appear in the Variables folder in the Project Organizer.

⬦ *If you rename a variable, Flare will automatically update all uses of the variable throughout your project.*

Creating a variable set

You can create multiple variable sets to organize your variables or to share company-wide variables between projects.

Shortcut	Tool Strip	Ribbon
Ctrl+T	📄 (Content Explorer	File > New

To create a variable set:

1 Select **File** > **New**.
   —OR—
   Right-click the **Variables** folder and select **Add Variable Set**. The Add File dialog box appears.

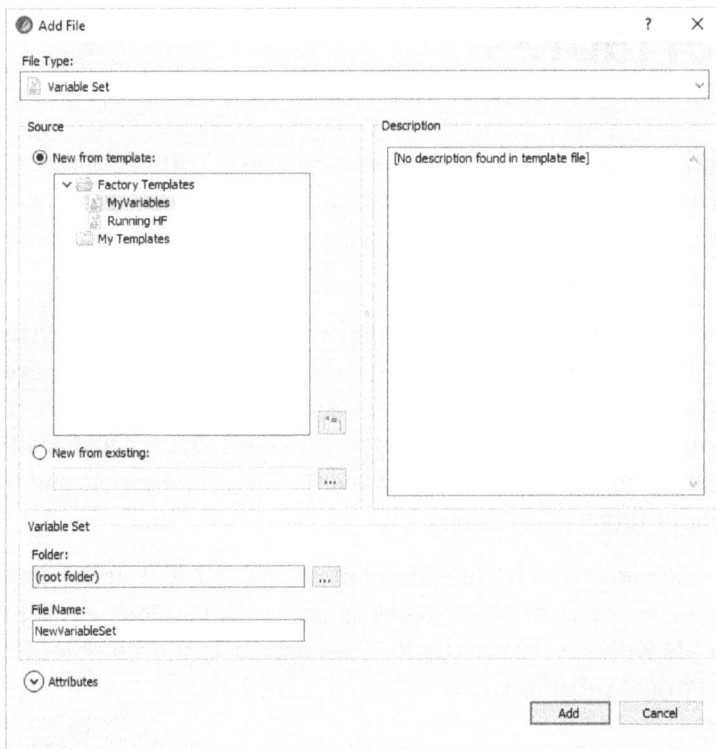

2 For **File Type**, select **Variable Set**.

3 Select a **Source** template.

4 Type a **File Name**.

5 Click **Add**.

Creating a variable

You can create as many variables as you need, and a variable can have multiple definitions. For example, a "GuideTitle" variable could have "User Guide," "Technical Specification," and "Quick Reference" definitions. Then, you could set the variable to use different definitions for a user guide, technical specification, and quick reference target.

To create a variable:

1 Double-click a variable set.
The Variable Set Editor window appears.

VariableSet Editor	Name ▲	Definition	Comment
▶	CompanyName	MadCap Software	
	PhoneNumber	858 123 4567	

2 Click 🔲 in the Variable Set Editor toolbar.

3 Type a name for the variable.

4 Type a definition for the variable.

5 If you want to add another definition:

   □ Click 🔲.

   □ Type the definition.

Creating a date/time variable

You can create a date/time variable to insert the date and/or time using one of Microsoft's date and time formats. For more information about Microsoft's date and time formats, see msdn.microsoft.com/en-us/library/8kb3ddd4.aspx

To create a date/time variable:

1 Double-click a variable set.
   The Variable Set Editor window appears.

VariableSet Editor	Name ▲	Definition	Comment
▶	CompanyName	MadCap Software	
	PhoneNumber	858 123 4567	

2 Click 🔲 in the Variable Set Editor toolbar.

3 Type a name for the variable.

4 Click inside the definition cell.
   The Edit Format dialog box appears.

5 Type a date/time format.
For example: ddd MM/dd/yy could appear as Sun 03/30/69

6 Select an **Update** option:

- **Manually** — displays the date/time when the variable was created or manually updated.

- **On Build** — displays the date/time when the target was built.

- **On File Creation** — displays the date/time when you created the file that contains the variable.

- **On File Save** — displays the date/time when you last saved the file that contains the variable.

- **On Project Save** — displays he date/time when you last saved all of the files in the project.

7 Click **OK**.

Inserting a variable

You can insert a variable into a topic, snippet, template page, or page layout. You can also insert a variable practically anywhere, including TOC items labels, glossary terms and definitions, and link destinations. To insert a variable in a dialog box, click ⧉.

Shortcut	Tool Strip	Ribbon
Ctrl+Shift+V	⧉ (XML Editor)	Insert > Variable

To insert a variable:

1 Position your cursor where you want to insert the variable.

2 Select **Insert** > **Variable**.
The Variables dialog box appears.

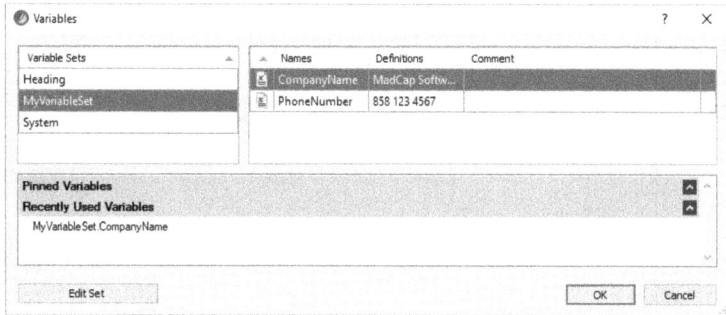

3 Select a variable set.

4 Select a variable.

5 Click **OK**.

TIP▶ *You can also drag a variable from the Project Organizer into the editor window.*

Overriding a variable's definition in a target

You can specify a variable's definition in a target to override the default value set in the variable set.

To override a variable's definition in a target:

1 Open the target.

2 Select the **Variables** tab.

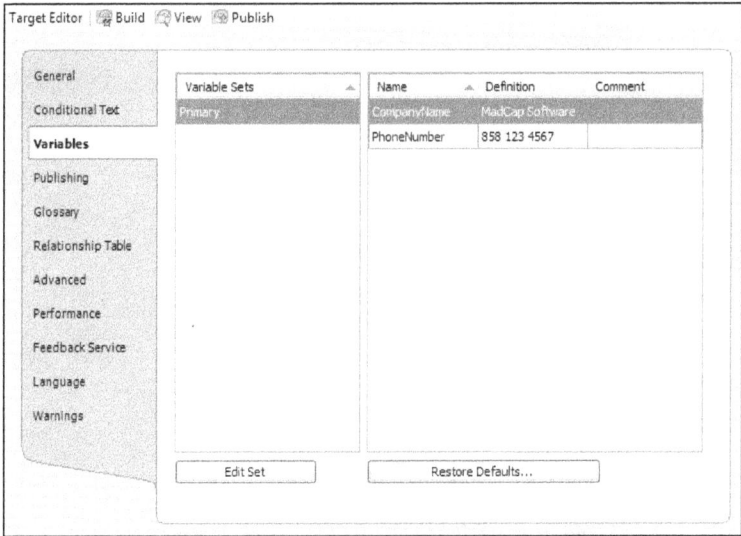

3   Click the variable's **Definition** cell and type a value. If you added multiple definitions when you created the variable, you can select a definition from the list.

Overriding a variable's definition in a topic

You can specify a variable's definition for every use of the variable within snippets in a topic. For example, you might use a "ProductName" variable in multiple snippets and insert these snippets into a topic. You can set the ProductName variable's definition for the topic and use the default definition (or another definition) for the ProductName variable in other topics.

To override a variable's definition in a topic:

1   Right-click a topic that contains snippets that include variables and select **Topic Properties**.

2   Select the **Variables** tab.

3   Select a **Variable Set**.

4   If the variable has multiple definitions, select a definition. If the variable only has one definition, double-click the variable and type a definition.

5   Click **OK**.

Overriding a variable's definition in a snippet

You can specify a variable's definition when it is used in an inserted snippet. For example, a snippet might contain a phone number that should change based on where the snippet is inserted.

To override a variable's definition in a snippet:

1   Open a topic.

2   Right-click a snippet that has been inserted into the topics and select **Snippet Variables**.
    The Snippet Variables dialog box appears.

3   Select a **Variable Set**.

4   If the variable has multiple definitions, select a definition. If the variable only has one definition, double-click the variable and type a definition.

5   Click **OK**.

Overriding a variable's definition in micro content

You can specify a variable's definition when it is used in a micro content block.

To override a variable's definition in a micro content block:

1   Open a micro content file.

2   Right-click a micro content phrase and select **Properties**.
    The Properties dialog box appears.

3 Select the **Micro Content Variables** tab.

4 If the variable has multiple definitions, select a definition. If the variable only has one definition, double-click the variable and type a definition.

5 Click **OK**.

Pinning a variable

You can pin frequently used variables in the Variables dialog box. Pinning a variable makes it easier to find.

To pin a variable in the variable list:

1 Select **Insert** > **Variable**.
The Variables dialog box appears.

2 In the **Recently Used Variables** box, hover over the variable.

3 Click ▰.
The variable is pinned.

Replacing text with a variable

You can use the Variables Suggestions report to find text and replace it with a variable.

To replace text with a variable:

1 Select **Analysis** > **Suggestions** > **Variable Suggestions**.
The Variable Suggestions report appears.

2 Highlight the item(s) you want to replace with a variable.

3 Click **Apply**.

Finding where variables are used in a project

You can use the Used Variables report to find where variables are in a project.

To find where variables are used in a project:

☐ Select **Analysis** > **Used Items** > **Used Variables**.
The Used Variables report appears.

Snippets

A snippet can include any type of content, including formatted text, links, images, tables, lists, and variables. I often create snippets for tables and steps that I need to use in multiple topics.

Snippets are stored in a file with a .flsnp extension. They appear in the Resources\Snippets folder in the Content Explorer.

Creating a snippet using existing content

You can select content within a topic and convert it to a snippet.

Shortcut	Tool Strip	Ribbon
Alt+H, S, P	Format > Create Snippet	Home > Create Snippet

To create a snippet using existing content:

1   Open the topic that contains the content you want to convert to a snippet.

2   Highlight the content you want to convert to a snippet.

3   Select **Home** > **Create Snippet**.
    The Create Snippet dialog box appears.

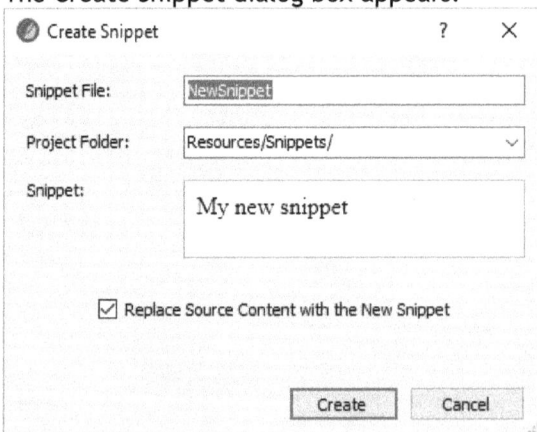

4   In the **Snippet File** field, type a new name for the snippet.

5 Leave the project folder selection as **Resources/Snippets.**

6 If you want the snippet to replace the highlighted text in the topic, select the **Replace Source Content with the New Snippet** option.

7 Click **Create.**
The snippet is created.

Creating a snippet using new content

You can also create a blank snippet and add content to it.

Shortcut	Tool Strip	Ribbon
Ctrl+T	🗒 (Content Explorer	File > New

To create a snippet using new content:

1 Select **File** > **New.**
The Add File dialog box appears.

2 For **File Type**, select **Snippet.**

3 Select a **Source** template.

4 Select a **Folder.**
By default, snippets are stored in the Resources/Snippets folder.

5 Type a **File Name** for the snippet.

6 Click **Add.**
The snippet appears in Content Explorer and opens in the XML Editor.

7 Click inside the snippet page in the XML Editor and add your content.

Inserting a snippet

You can insert snippets into topics, template pages, and page layouts.

Shortcut	Tool Strip	Ribbon
Ctrl+R	(XML Editor)	Insert > Snippet

To insert a snippet to a topic:

1 Position the cursor where you want to insert the snippet.

2 Select **Insert** > **Snippet**.
   The Insert Snippet Link dialog box appears.

3 Select a snippet.
   A preview of the snippet appears.

4 Click **OK**.
   The snippet is added to the topic.

Pinning a snippet

You can pin frequently used snippets in the Insert Snippet Link dialog box. Pinning a snippet makes it easier to find.

To pin a snippet in the snippet list:

1 Select **Insert** > **Snippet**.
   The Insert Snippet Link dialog box appears.

2 In the **Recently Used Snippets** box, hover over the snippet.

3 Click ▬.

The snippet is pinned.

Replacing content with a snippet

You can use the Snippets Suggestions report to find content and replace it with a snippet.

To replace content with a snippet:

1 Select **Analysis** > **Suggestions** > **Snippet Suggestions**.
The Snippet Suggestions report appears.

2 Highlight the item(s) you want to replace with a snippet.

3 Click **Apply**.

Finding where snippets are used in a project

You can create an information report to find where snippets are used in a project.

To find where one snippet is used in a project:

☐ Right-click the snippet and select **View Links**.
The Link Viewer appears.

To find where all snippets are used in a project:

1 Select **File** > **New**.
—OR—
Right-click the **Reports** folder and select **Add Report File**.

The Add File dialog box appears.

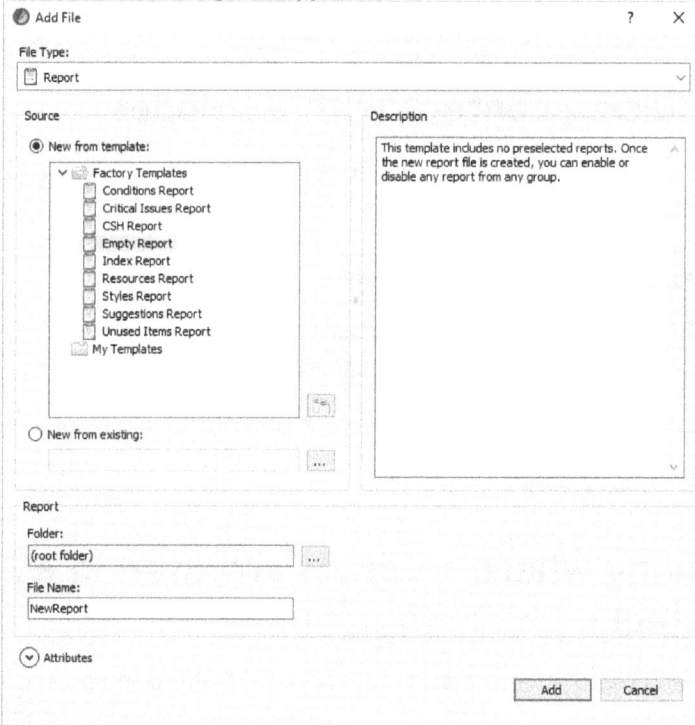

2 For **File Type**, select **Report File**.

3 Select a **Source** template.

4 Type a **File Name**.

5 Click **Add**.
   The report appears in the Reports folder in the Project
   Organizer and opens in the Report Editor.

6 In the **Snippets** group, select **Used Snippets**.

7 Click **Generate**.

Condition tags

Condition tags can be used to include or exclude content when you create a target. For example, you could create an "InternalOnly" tag and exclude it when you create a knowledgebase for customers.

You can apply a condition tag to almost anything: folders, topics, content in topics, TOC entries, index entries, glossary terms, stylesheets, variables, snippets, and micro content blocks. You can even apply a condition tag to content using a style. See "Applying condition tags with a style" on page 241.

Condition tags are stored in a condition tag set. When you import a RoboHelp project or FrameMaker document that contains conditional build tags, your tags are stored in a condition tag named after your import file.

TIP *If you insert an image created with MadCap Capture or insert a video created with MadCap Mimic into a Flare project, you can use your Flare project's condition tags in the image or video.*

Conditional tag sets have a .flcts extension. They appear in the Conditional Text folder in the Project Organizer.

'What are those boxes in the Content Explorer?'

They're called "condition tag boxes."

When you create a condition tag, you assign a color to the tag. This color is used to identify tagged content. If a folder or topic is associated with a tag, the condition tag box is filled with the tag's assigned color. If the folder or topic is associated with multiple tags, the condition tag box uses vertical stripes to show each tag's color.

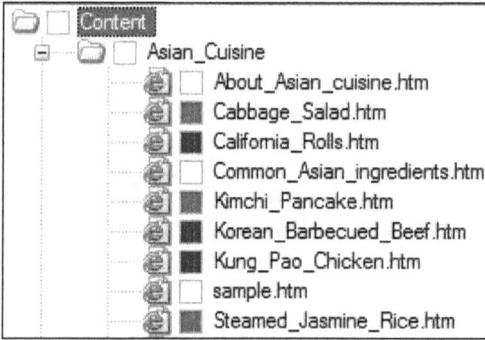

Creating a condition tag

Shortcut	Tool Strip	Ribbon
Ctrl+T	📄 (Content Explorer	File > New

To create a condition tag:

1 Open the Project Organizer.

2 Open the **Conditional Text** folder.

3 Double-click a condition tag set.
The Condition Tag Set Editor appears.

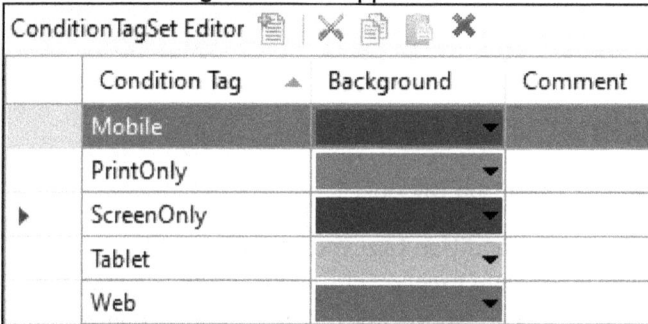

4 Click 📄 in the Condition Tag Set Editor toolbar.
A new tag appears.

5 Double-click the new tag's name.

6 Type a new name for the tag and press **Enter**.

7 Select a color.

Applying a tag to content

To apply a condition tag to content:

1 Open a topic.

2 Select the content to be tagged.
You can apply a tag to any content, including characters, words, paragraphs, table columns/rows, items in a list, and images.

3 Select **Home** > **Conditions**.
—OR—
Press **Ctrl+Shift+C**.
The Condition Tags dialog box appears.

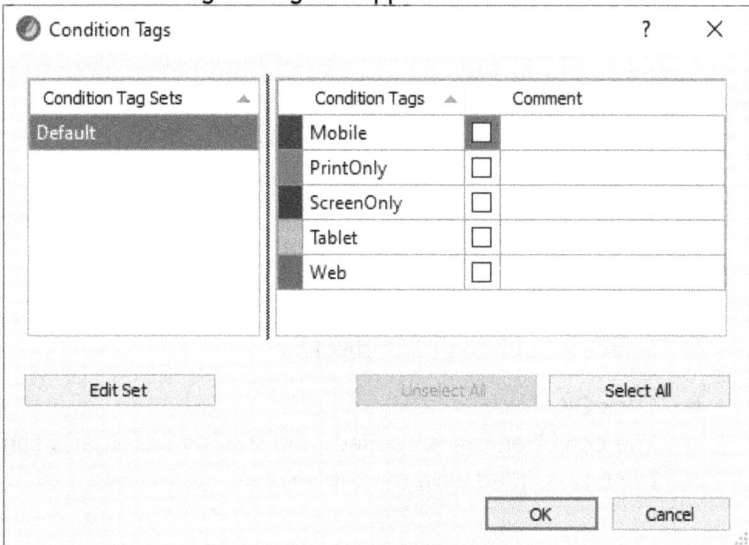

4 Select a condition tag's checkbox.

5 Click **OK**.
The condition tag is applied. The tagged content is shaded using the tag's color.

Applying a tag to a topic, file, or folder

To apply a condition tag to a topic, file, or folder:

1 Open the Content Explorer.

2 Select the topic, file, or folder to be tagged.

3 Click 🖼 in the Content Explorer toolbar.
The Properties dialog box appears.

4 Select the **Conditional Text** tab.

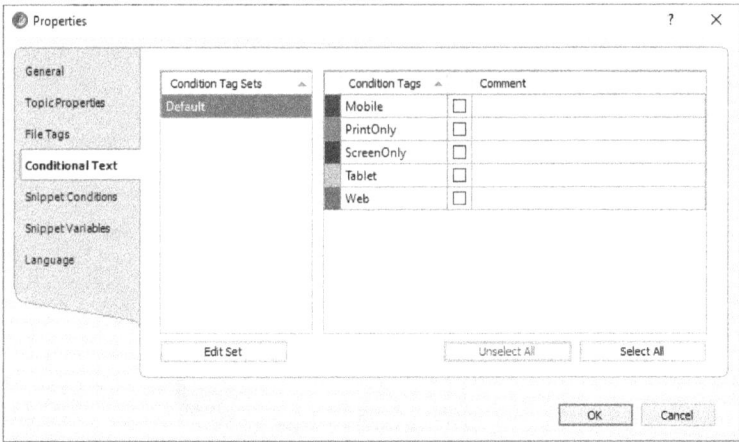

5 Select a condition tag's checkbox.

6 Click **OK**.
The condition tag is applied, and the file or folder's condition tag box is filled with the tag's color.

To apply a condition tag when you create a topic:

1 Select **File > New**.
The Add File dialog box appears.

2 Select a **File Type**, **Template**, and **Folder**.

3 Type a **File Name**.

4 Click **Attributes**.

5 For **Condition Tags**, click the ... (browse) link.
The Condition Tags dialog box appears.

6 Select a **Condition Tag Set**.

7 Select a **Condition Tag** (or tags) and click **OK**.

Applying a tag to a TOC book or page

You can also apply a condition tag to a TOC book or page instead of a topic. For online targets, excluding a tagged TOC item only removes it from the TOC. Excluding a tagged topic removes the topic from the TOC, index, and search.

To apply a condition tag to a TOC book or page:

1 Open the Project Organizer.

2 Double-click a TOC.

3 Select a book or page and click 📄 in the TOC Editor.
The Properties dialog box appears.

4 Select the **Conditional Text** tab.

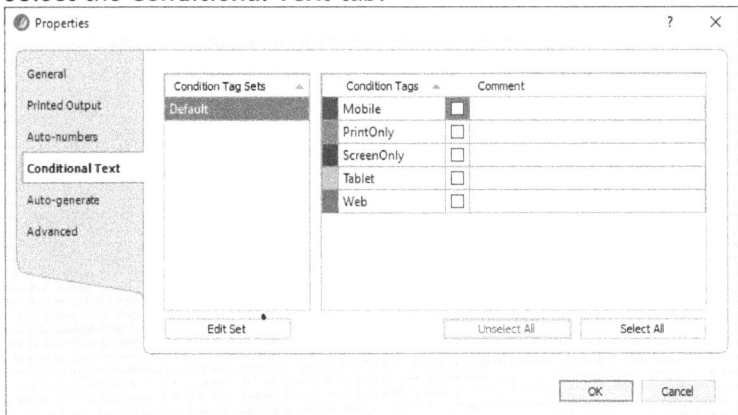

5 Select a condition tag's checkbox.

6 Click **OK**.
The condition tag is applied, and the TOC book or page's condition tag box is filled with the tag's color.

Applying a tag to an index entry

To apply a condition tag to an index entry:

1 Open a topic.

2 Select an index entry marker.

3 Select **Home** > **Conditions**.
The Condition Tags dialog box appears.

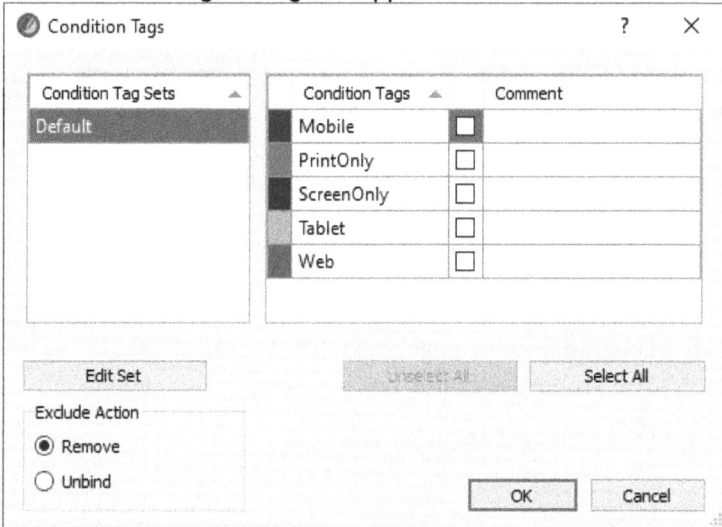

4 Select a condition tag's checkbox.

5 Click **OK**.
The condition tag is applied. A condition tag box appears inside the tagged index entry's marker.

Applying a tag to a glossary term

To apply a condition tag to a glossary term:

1 Open a glossary.

2 Double-click a glossary term.
The Properties dialog box appears.

3 Select the **Conditional Text** tab.

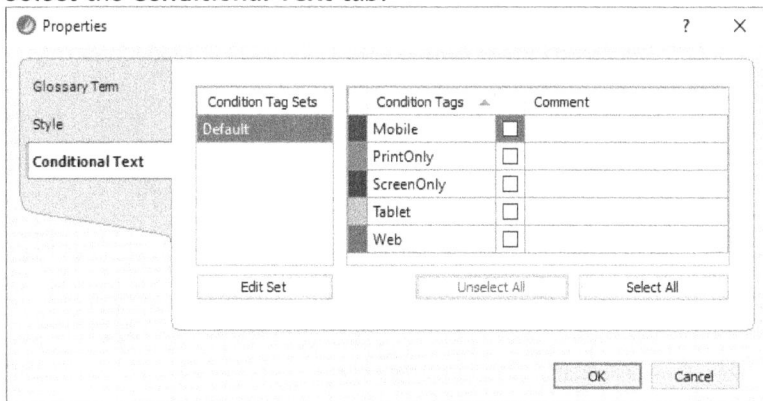

4 Select a condition tag's checkbox.

5 Click **OK**.
The condition tag is applied.

Using snippet conditions

You can use snippet conditions to reuse content that needs to be slightly different when it is used in certain topics. For example, you might need to reuse a procedure in ten topics, but in three of the topics you might not need to include one step. You can apply a condition tag to that step and include it in seven topics and exclude it in the other three topics.

To use snippet conditions:

1 Create the snippet.

2 Apply the condition tag to the content in the snippet.

3 Insert the snippet into your topics.

4 Select **View** > **File List**.

5 Select the topics that should include or exclude the tagged content in the snippet.

6 Right-click the selected topics and select **Properties**.
The Properties dialog box appears.

7 Select the **Snippet Conditions** tab.

8 Set whether the tag should be included or excluded.

9 Click **OK**.

Using micro content conditions

You can use micro content conditions to reuse content that needs to be slightly different when it is used in micro content blocks. For example, you might want to reuse a procedure in a micro content block but exclude the notes, tips, and feedback statements inside the steps. You can apply a condition tag to the notes, tips, and feedback statements and exclude them in your micro content block.

To use micro content conditions:

1 Open the topic that contains the micro content block.

2 Apply the condition tag(s) to the content in the micro content block.

3 Open a micro content file.

4 Right-click a phrase and select **Properties**.
The Properties dialog box appears.

5 Select the **Micro Content Conditions** tab.

6 Set whether each tag should be included or excluded.

7 Click **OK**.

Reviewing how conditions are applied in a project

You can use the Used Conditions and Files with Snippet Conditions reports to review how conditions are applied in a project.

To review how conditions are applied in a project:

☐ Select **Analysis** > **Used Items** > **Used Condition Tags**.
The Used Condition Tags report appears.

To review how snippet conditions are applied in a project:

☐ Select **Analysis** > **More Reports** > **Files with Snippet Conditions**.
The Files with Snippet Conditions report appears.

Including or excluding a tag when you build a target

By default, all of your content is included when you build a target. In fact, tagged content is also included unless you exclude it.

For example, a project might contain two condition tags: A and B. Some content is tagged as A, some content is tagged as B, and some content is tagged as A and B.

To include...	and exclude...	Use these settings
Everything	nothing	None (this is the default)
☐ Untagged content ☐ A	B	Include A —OR— Exclude B ◇ *If you include a tag, the default changes to exclude. If you have several tags that you need to exclude and one tag you need to include, it will be easier to Include A.*
Untagged content	☐ A ☐ B	☐ Exclude A ☐ Exclude B
☐ Untagged content ☐ A ☐ A + B (content with both tags)	B	Include A ◇ *If you include A, it will be included even if the content is also tagged as B.*
☐ Untagged content ☐ A	A + B (content with both tags) B	Use an Advanced Expression: include[Tagset.A and not Tagset.B]

▣ *You can use responsive condition tags to show or hide content based on the width of the screen. See "Using responsive condition tags" on page 293.*

To include or exclude a tag when you build a target:

1 Open a target.

2 Select the **Conditional Text** tab.

3 Select **Basic**.

4 Select the tag(s) you want to **Exclude**.

5 If you have assigned more than one tag to topics or blocks on content in topics, select the tag(s) you want to **Include**.

To create an advanced build expression:

1 Open a target.

2 Select the **Conditional Text** tab.

3 Select **Advanced**.

4 In the **Advanced** text box, type a build expression.

Micro content

Micro content is composed of a phrase and a response. Many micro content blocks are question and answer, but you can also create micro content to define a term or provide steps for a procedure. Micro content should be a relatively short, focused, self-contained "chunk" of easy to read and understand information.

Micro content is most often used in featured snippets and knowledge panels in search results. However, micro content can be used for field-level help and integrated into chatbot, voice-activated search, AR, and VR systems. Micro content is stored in a micro content file with a .flmco extension. They appear in the Resources\Micro Content folder in the Content Explorer.

Creating a micro content file

You can store multiple micro content blocks in a micro content file. Or, you can organize them into separate micro content files. When you setup featured snippets and the knowledge panel in Flare's search results, you can specify whether they include micro content blocks from all or a specific micro content file.

Shortcut	Tool Strip	Ribbon
Ctrl+T	📝 (Content Explorer	File > New

To create a micro content file:

1   Select **File** > **New**.
    The Add File dialog box appears.

2   For **File Type**, select **Micro Content**.

3   Select a **Source** template.

4   Type a **File Name**.

5   Click **Add**. The micro content file opens in the Micro Content Editor.

Creating micro content

You can create micro content blocks directly to a micro content file. This approach is often more efficient if you are adding multiple micro content blocks or if you plan to add alternate search phrases.

To create a micro content block:

1 Open a micro content file.
   By default micro content files are stored in the Resources folder in the Content Explorer.

2 Click .

3 Type a search phrase and press **Enter**.

4 If you want to add an alternate phrase, click .

5 Click inside the Response Editor on the right and add your content.
   —OR—
   Click to the right of the main phrase and click **Add Link** to use a topic or snippet's content as the response.

Creating micro content using existing content

You can select content within a topic and convert it to micro content.

Shortcut	Tool Strip	Ribbon
Alt+H, M, T	Format > Create Micro Content	Home > Create Micro Content

To create micro content using existing content:

1 Open the topic or snippet that contains the content you want to convert to micro content.

2 Highlight the content you want to convert to a micro content.

3 Select **Home** > **Create Micro Content**.

The Create Micro Content dialog box appears.

4 Type a search **Phrase**.

5 Select a **Micro Content File**.

6 Click **OK**.

The micro content block is created.

Adding a micro content proxy

You can add three types of micro content proxies to topics, template pages, or a search results topic:

- ☐ **FAQ proxy** — an alphabetized, static list of question-and-answer micro content blocks.

- ☐ **Knowledge proxy** — dynamic micro content blocks that provide additional or links to related information based on the user's search term or the current topic. Examples: topic author, related topics

- ☐ **Promotion proxy** — dynamic and randomly selected micro content blocks that are relevant to the user's search term or the current topic. Examples: new features or upcoming events

To add a micro content proxy:

1 Open a topic or template page.

2 Position the cursor where you want to add the micro content proxy.

3 Select **Insert** > **Proxy** and select either **FAQ Proxy**, **Knowledge Proxy**, or **Promotion Proxy**.

4 Type a **Proxy Title**.

5 If you have created an FAQ, Knowledge, or Promotion skin component, select the skin.

6 If you have created a separate stylesheet to format your micro content blocks, select **Allow micro content stylesheets**.

7 Select which micro content files to use to find micro content block matches.

8 If you are adding a knowledge or promotion proxy:

   □ Select how the micro content should be included.

   □ Select a maximum number of micro content results.

9 If you are adding an FAQ proxy:

   □ Select whether the micro content blocks should be alphabetized.

   □ Select **Generate FAQ structured data** if you want your FAQ blocks to be tagged as FAQs for Google's search results.

10 Click **OK**.

Sample questions for this section

1 Where are variables stored?
A) In variable sets in the Project Organizer
B) In variable files in the Content Explorer's Resources folder
C) In variable files in the Content Explorer's Variables folder
D) Inside topics

2 A variable's definition can be set: (select all that apply)
A) For a topic
B) In the VariableSet Editor
C) In a target on the Variables tab
D) In the Project Properties on the Variables tab

3 How many definitions can you specify for a variable?
A) One
B) Two
C) Up to ten
D) As many as you need.

4 A snippet can contain: (select all that apply)
A) Formatted text
B) Tables
C) Lists
D) Variables

5 Where are snippets stored?
A) In snippet sets in the Project Organizer
B) In snippet files in the Project Organizer
C) In snippet files in the Content Explorer
D) Inside topics

6 A condition tag can be applied to: (select all that apply)
A) Topics
B) Folders
C) TOC books and pages
D) Index keywords

7 What happens if you rename a condition tag after it has been applied to content?
A) Flare automatically updates the content to use the new condition

tag name.

B) Flare asks if you want to update the content to use the new name.

C) The condition tag is removed from your content.

D) Nothing—you must update the content yourself to use the new name.

8   Condition tags can be applied to the following targets:

A) PDF

B) Word

C) HTML Help

D) Any type of target

Build and publish

This section covers:

- Targets
- Context-sensitive help

Targets

You can create eleven types of online and print targets from Flare. Each target type is summarized in the tables below.

Online targets

Target	Description
Clean XHTML	An online format that can be used to embed your output into other systems such as an application, a wiki, an eLearning system, or Salesforce.
Eclipse Help	An online format that can be used to create an Eclipse Help plug-in that can be viewed in the Eclipse Help viewer.
HTML Help	An online format developed by Microsoft that can be used to create Windows-based help. Microsoft no longer maintains or supports the HTML Help format, and they have promised an eventual replacement. HTML Help targets run in the HTML Help Viewer and they have a .chm (often pronounced "chum") file extension.
HTML5	An online format developed by MadCap Software to replace WebHelp. HTML5 provides numerous skin design options, and the skins can be highly customized. HTML5 also provides three search options: MadCap Search, Google Search, and Elasticsearch.
WebHelp	An older online format developed by MadCap Software that can be used to create Web-based content such as help systems. WebHelp runs in a browser, so it is not Windows-specific like HTML Help.
WebHelp Plus	A special version of WebHelp that provides faster searches and allows users to search content in .pdf, .doc, and .xls files. Unlike WebHelp, WebHelp Plus requires a Microsoft Web server with IIS.

'What is Clean XHTML?'

Clean XHTML is a format created by MadCap that can be used to integrate your output into another system, such as a wiki or Salesforce. Clean XHTML outputs don't include skins, search, breadcrumbs, or menus, and Clean XHTML topics don't include any MadCap-specific tags or JavaScript files. Since Clean XHTML topics don't include MadCap's scripts, they do not include drop downs, togglers, popups, or any other MadCap-specific features.

'What is Eclipse Help?'

Eclipse Help is a format that can be used to create an Eclipse Help plug-in. Eclipse Help plug-ins can be viewed in the Eclipse Help viewer. For information about customizing the Eclipse Help viewer, see help.eclipse.org/kepler/index.jsp.

'What is HTML5?'

HTML5 is a recommendation from the W3C that was created to replace HTML and XHTML. MadCap's HTML5 format uses HTML5, and it supports the most recent features in HTML and CSS. It is (by far) the most commonly used online format.

'What about WinHelp?'

Flare cannot create WinHelp. Microsoft's WinHelp is an old help format that was replaced by HTML Help in the late 1990s. Microsoft slowly phased out WinHelp over ten years, and they stopped supporting it with Windows Vista.

Print targets

Target	Description
EPUB	A print format that is designed to be viewed online using a phone, tablet, or eReader.
MOBI	A print format that is designed to be viewed on a Kindle.
PDF document	A print format developed by Adobe. The PDF format combines all of your topics and images into one file with a .pdf ('Portable Document Format') extension.
Word document	A print format that creates Microsoft .doc or .docx documents.

'What is EPUB?'

EPUB is an online book format developed by the International Digital Publishing Forum (IDPF). It was designed and developed to format books for reading on electronic devices.

'What is MOBI?"

MOBI is an online book format that is owned by Amazon. It is used to create content for the Kindle. If you build an EPUB target, you can also build a MOBI version.

Creating a target

You should create a different target for each version of your online or print documents. For example, if you need to create online help and PDFs for the "Standard" and "Professional" versions of your product, you should create four targets.

Shortcut	Tool Strip	Ribbon
Ctrl+T	📄 (Content Explorer	File > New

To create a target:

1 Select **File** > **New**.
   —OR—
   Right-click the **Targets** folder and select **Add Target**.

   The Add File dialog box appears.

2　For **File Type**, select **Target**.

3　Select a **Source** template.

4　Type a **File Name**.
You don't have to type the .fltar extension. Flare will add it for you if you leave it out.

5　Select an **Output Type**.

6　Click **Add**.
The target appears in the Targets folder in the Project Organizer and opens in the Target Editor.

Specifying the primary target

The primary target is the target that you plan to create the most often. When you preview a topic, the topic will appear as it will appear in the primary target.

To specify the primary target:

☐ Right-click a target and select **Make Primary**.

Setting up an online target

You can specify the skin, template page, TOC, condition tags, variables, and glossaries that are used in your online target.

To set up an online target:

1 Open a target.

2 On the **General** tab, select the following options:

☐ **Output Type** — the help format that you are creating.

☐ **Comment** — a short description of the target.

☐ **Startup Topic** — the first topic that appears when the user opens your help system.

☐ **Skin** — the skin file specifies the size, appearance, and features included in a target.

☐ **Primary TOC** — the TOC that will be used for the target.

☐ **Browse Sequence** — the browse sequence that will be used for the target.

☐ **Primary Stylesheet** — the stylesheet that will be used for all topics. Setting the primary stylesheet will override stylesheets that have been applied to specific topics.

☐ **Output File** — the name of the main entry (or 'start') file for your help system.

☐ **Output Folder** — where the generated help files will be created.

☐ **Convert stylesheet styles to inline styles** — converts your stylesheet formatting to inline formatting to avoid using external stylesheet files (Clean XHTML only)

3 Select the **Skin** tab.

4 Select a **Skin**.

5 If you are using skin components, select your skin component(s).

6 Select the **Conditional Text** tab.

7 Select whether you want to include or exclude each tag. See "Including or excluding a tag when you build a target" on page 318.

8 Select the **Variables** tab.

9 If you need to change a variable's definition for the target, click inside its **Definition** cell and type a new value.

10 If you want to publish your target to a network drive or web server:

  □ Select the **Publishing** tab.

  □ Select a destination.
    —OR—
    Click **New Destination** to create a publishing destination.

11 Select the **Glossary** tab and select the following options:

  □ **Glossary Term Conversion** method — how your glossary terms are converted in your topics.

  □ **Glossary** — the glossary (or glossaries) that are included in your target.

12 If you are using relationship tables and links, select the **Relationship Table** tab and select the relationship tables to use in your target.

13 Select the **Search** tab.

14 Select a **Search Engine**.
    Google Search and Elasticsearch are not available for tripane skins.

  MadCap Search — works locally or on a server. The result ranking considers heading levels, index keywords, and glossary links, and glossary terms can appear at the top of the results.

Google Search — requires your content to be publicly available, and the search results will include ads. However, it does provide "fuzzy matching" for incorrectly spelled terms, and the search results can include content from other sites and/or folders. Google Search is not available for tripane targets.

Elasticsearch — works locally or on a server. It provides numerous features, including fuzzy matching, predictive search, and autocompletion. Elasticsearch requires a Java Development Kit (JDK). Since Oracle charges for their JDK, MadCap has added support for the free OpenJDK. You can download it at madcapsoftware.com/downloads/java.aspx

15 Select **Generate Sitemap** if your content can be accessed by search engines. A sitemap will improve search engine optimization (SEO).

16 Select the **Advanced** tab and select the options you would like to use. Commonly set options include:

□ **Content to Include** - select whether you want to include all content (except content you are excluding using condition tags), content linked directly or indirectly by the topics you are including in your target, or only topics included in your TOC.

□ **Template Page** — select a template page to apply to your topics.

17 If you are building a large (8,000+ topic) online target, select the **Performance** tab and set the following options:

□ **Condense JavaScript files** — condensing the JavaScript files into one file can improve server performance if you have a lot of users.

□ **Index** — pre-merging the index file and using smaller index "chunks" can make the index open faster for the user.

□ **TOC** — using smaller TOC chunks can make the TOC open faster for the user.

□ **Search Database** — you can exclude non-words from search (such as code examples), pre-merge or use smaller search chunks to make the search open faster, and/or use larger n-grams to make a Japanese, Chinese, or Korean language search more accurate.

18 If you need to run a script before or after you build your target, select the **Build Events** tab and add your script(s). For example, you could create a script to copy a file into your output folder after you build.

19 If you are using MadCap Central or Pulse and want to gather usage data, select the **Analytics** tab and type your license and key.

20 Select the **Language** tab and select a language for the skin.

21 Select the **Warnings** tab to enable or disable build warnings. Build warnings should normally be enabled, but you might temporarily disable some messages if they aren't important or don't apply to your target.

22 Click **Save**.

Setting up a print target

You can create a print target to specify the condition tags, variables, and glossaries that are used in your print documentation.

To set up a print target:

1 Open a target.

2 On the **General** tab, select the following options:

□ **Output Type** — the help format that you are creating.

□ **Comment** — a short description of the target.

□ **Primary TOC** — the TOC that will be used for the target.

□ **Primary Page Layout** — the page layout that will be used for all topics in a print target. Setting the template page layout will override page layouts that have been applied to specific topics.

- Primary Stylesheet — the stylesheet that will be used for all topics. Setting the primary stylesheet will override stylesheets that have been applied to specific topics.

- Output File — the name of the generated document.

- Output Folder — where the generated files will be created.

3 Select the Conditional Text tab.

4 Select whether you want to include or exclude each tag. See "Including or excluding a tag when you build a target" on page 318.

5 Select the Variables tab.

6 If you need to change a variable's definition for the target, click inside its Definition cell and type a new value.

7 If you want to publish your print document to a file or web server:

- Select the Publishing tab.

- Select a destination.
  —OR—
  Click New Destination to create a publishing destination.

8 Select the Glossary tab and select a Glossary (or glossaries) to include in your documentation.

9 If you are using relationship tables and links, select the Relationship Table tab and select the relationship tables to use in your target.

10 Select the Advanced tab and select the options you would like to use. Commonly set options include:

- Generate TOC, Index, or Glossary Proxy — select each option to automatically include a TOC at the beginning and/or an index and/or glossary at the end in your document.

  ◇ If the first topic in your TOC is set to use the "title" page type in your page layout, it will appear before the

TOC. If not, the TOC will appear first. See "Applying a page layout to a topic" on page 270.

- **Preserve Tracked Changes** — if you have tracked changes, you can include them in Word and PDF targets.

- **Stylesheet Medium** — if your stylesheet contains a medium (most people use the "print" medium), you can use the medium's styles or the stylesheet's default styles.

- **Expanding Text Effects** — select how expanding text should appear in your print documents.

- **Text Popup Effects** — select how text popups should appear in your print document.

- **Generated TOC** — select whether your heading levels should match your TOC or the heading tags you used in your topics. Also, select whether you want to create headings for unlinked books in your TOC.

- **Multi-Document Native PDF Output** — if you are creating a PDF document, select this option to create multiple documents based on chapter breaks in the TOC. See "Specifying Chapter Breaks" on page 194.

- **Redacted Text** — if you are creating a PDF document and using the redacted style property, you can set how redacted text will appear.

11 If you need to run a script before or after you build your target, select the **Build Events** tab and add your script(s).

12 If you are creating a PDF target, select the **PDF Options** tab and set the following options:

- **Image Compression** — set the **Compression** option to **Automatic** to use Flare's lossless compression for non-JPG images or select **JPG** to convert all images to JPGs with some compression.

- **Document Properties** — type a title, author, subject, and any keywords you want to include and select whether you want to include crop and registration marks or convert RGB colors to CMYK.

- **Copyright** - select a copyright status and type a copyright notice and information URL.

- **Generate Tagged Document for PDF/UA** — specify whether you want to include tags that are used by accessibility applications such as screen readers.

- **Initial View** — select the magnification level, whether the Bookmarks panel should appear, the page layout, and the text for the title bar.

- **Security** — specify whether you want to require a password or if you want to restrict the user from printing, editing, or copying text and graphics.

13 If you are creating a Word target, select the **MS Word Options** tab and set the following options:

- **Document Properties** — type a title, author, subject, and any keywords you want to include.

- **Generate Multiple Documents for MS Word Output** — if you are creating a Word document, select this option to create multiple documents based on chapter breaks in the TOC. See "Specifying Chapter Breaks" on page 194.

14 If you are creating an EPUB target, select the **EPUB Options** tab and set the following options:

- **Meta data** — type a title, author, publisher, and any other document properties you want to include.

- **Cover page** — if you want to add a cover page, select a cover page image.

- **Validate Epub 3 output** — specify whether you want to validate the EPUB document using the free EpubCheck application.

- **Generate MOBI output** — specify whether you want to create a MOBI document based on your EPUB document using Amazon's free KindleGen application.

- **Optimization settings** — specify whether you want to convert MathXML equations to PNG images, embed fonts in your EPUB, or enable dynamic content such as drop-downs, help controls, and slideshows.

◇ *Enabling dynamic content may make your dynamic content appear distorted in some EPUB readers.*

15 Select the **Warnings** tab to enable or disable build warnings. Build warnings should normally be enabled, but you might temporarily disable some messages if they aren't important or don't apply to your target.

16 Click **Save.**

Building a target

By default, your generated targets are stored in a folder named "Output." You can change the default folder, but most users use the default folder.

Shortcut	Tool Strip	Ribbon
F6	🖼 (Review toolbar)	Project > Build Primary

To build a target:

1 Select the **Project** ribbon, click the down arrow beside the **Build Primary** button, and select a target.
—OR—
Right-click a target and select **Build.**

2 If you made any changes to the target, Flare will prompt you to save your changes. Click **Yes.**
The Builds pane appears. Flare will build the target in the background, so you can continue working in your project.

3 When the build is complete, highlight the target in the Builds pane and click **View Output.**
—OR—
Right-click the target and select **View.**

4 Select a browser, "smartphone" or "tablet."

If you select smartphone or tablet, the content will simply open in your default browser in a window that is sized to match a smartphone or tablet. It will not open in a smartphone or tablet emulator.

*You can select **Project** > **Clean Project** to delete everything in the Output folder.*

Building a target from the command line

Flare targets can be compiled from the command line using the MadBuild command.

To build a target from the command line:

1 Open a command prompt.

2 Navigate to the directory where you installed Flare. The default directory for Flare 2023 is program files\madcap software\madcap flare 19\flare.app.

3 Type **madbuild -project <path><projectname> -target <targetname>**. For example:

    madbuild –project c:\myFolder\myProject.flprj
    –target myHTML5

If any value contains a space, it must be enclosed in quotation marks. For example: "c:\my Folder\my Project.flprj"

To build all of your targets from the command line:

1 Open a command prompt.

2 Navigate to the directory where you installed Flare. The default directory for Flare 2023 is program files\madcap software\madcap flare 19\flare.app.

3 Type **madbuild -project <path><projectfilename>**. For example:

    madbuild –project c:\myFolder\myProject.flprj

Building an HTML5 target to Central from the command line

If your project is bound to Central, you can build and publish your HTML5 targets to Central from the command line.

To build an HTML5 target and publish it to Central from the command line:

1 Open a command prompt.

2 Navigate to the directory where you installed Flare. The default directory for Flare 2023 is program files\madcap software\madcap flare 19\flare.app.

3 Type **madbuild -project <path><projectfilename> -centralUsername <email address> -centralPassword <password> -target <targetname>**. For example:
```
madbuild –project c:\myFolder\myProject.flprj
–target myHTML5Target -centralUsername
scott@clickstart.net -centralPassword P@ssw0rd
```

Building a target from the command line with credentials

By default, the MadBuild command will use the credentials stored in the Windows registry. You can include publishing credentials, such as a user name and password, if you want to use a different set of credentials.

Credentials can be included for the following destination types:

- MadCap Central
- FTP
- Salesforce
- ServiceNow
- SFTP
- Zendesk

To build a target from the command line with credentials:

1 Open a command prompt.

2 Navigate to the directory where you installed Flare.
The default directory for Flare 2023 is
program files\madcap software\madcap flare 19\flare.app.

3 Type **madbuild -project <path><projectfilename> -target
<targetname> -credentials <credential URL format>**.

4 Add the credential URL format based on your destination type:

MadCap Central
```
"{\"https://portal.madcapcentral.com/\":
{\"username\":\"scott@example.com\", \"password\":
\"MyPassword\"} }"
```

FTP
```
"{\"https://docs.mysite.com/MyTarget\":
{\"username\":\"sdeloach\", \"password\":
\"MyPassword\"} }"
```

Salesforce
```
"{\"https://login.salesforce.com/services/oauth2/
authorize\": {\"username\":\"scott@example.com\",
\"password\": \"MyPassword"} }"
```

ServiceNow
```
"{\"https://mysite.service-now.com/oauth_token.do\":
{\"username\":\"scott.deloach\", \"password\":
\"MyPassword\"} }"
```

SFTP
```
"{\"https://myappserver:0101/MyTarget\":
{\"username\":\"sdeloach\", \"password\":
\"MyPassword\"} }"
```

Zendesk
```
"{\"https://mysite.zendesk.com/oauth/authorizations
/new\": {\"username\":\"scott@example.com\",
\"password\": \"MyPassword\"} }"
```

Building a target from the command line with credentials file

You can store the credentials in a JSON file rather than including them in the MadBuild command directly.

To build a target from the command line with credentials:

1 Create a JSON file to store the credentials.

2 Setup the JSON file using the following format:

```
{
    "https://mysite.com": {
    "username": "myusername",
    "password": "mypassword
}
```

3 Navigate to the directory where you installed Flare.
The default directory for Flare 2023 is
program files\madcap software\madcap flare 19\flare.app.

4 Type **madbuild -project <path><projectfilename> -target <targetname> -credentials-file <path to credentials file>**. For example: `madbuild -project c:\myFolder\myProject.flprj -target myHTML5Target -credentials-file "C:\MyCredentialsFolder\MyCredentialsFile.json"`

Adding a build event

You can use build events to run command line events before and/or after a target is generated. For example, you could use a build event to create a folder and copy your output to the folder after it is built.

◇ *Pre-build events run while the build is building and do not pause the build.*

To add a build event to a target:

1 Open a target.

2 Select the **Build Events** tab.

3   In the **Pre-Build Event Command** and/or **Post-Build Event Command** text box, type your build event.
    For example:
    `C:/create_folder.bat`

4   If you want to include a variable in your build event, click **Insert Build Variable**.

    You can use the following built-in variables in a build event:
    □   TargetName
    □   ProjectName
    □   ProjectDirectory
    □   OutputDirectory

Saving the build log

After you build a target in Flare, Flare will automatically save the build log. The build log contains a list of any build messages and errors, and it can be used to troubleshoot problems with your project. You can select where Flare saves the build log.

To select a destination for the build log:

1   Select **File** > **Options**.
    The Options dialog box appears.

2   Select the **Build** tab.

3   Select a **Log Destination**.

4   If you want to include timestamps, select **Use timestamps in log file**.

5   Click **OK**.

Viewing a target

Shortcut	Tool Strip	Ribbon
Shift+F6	(Review toolbar)	Project > View Primary

To view a target:

☐ Select the **Project** ribbon, click the down arrow beside the
View Primary button, and select a target.
—OR—
Right-click a target and select **View**.

Publishing a target to a file server

You can publish any target to a file server. For example, you could
publish your targets to an "archive" server after each release.

Shortcut	Tool Strip	Ribbon
Ctrl+T	📄 (Content Explorer	File > New

To create a publishing destination for a file server:

1 Select **File** > **New**.
—OR—
Right-click the **Destinations** folder and select **Add Destination**.

The Add File dialog box appears.

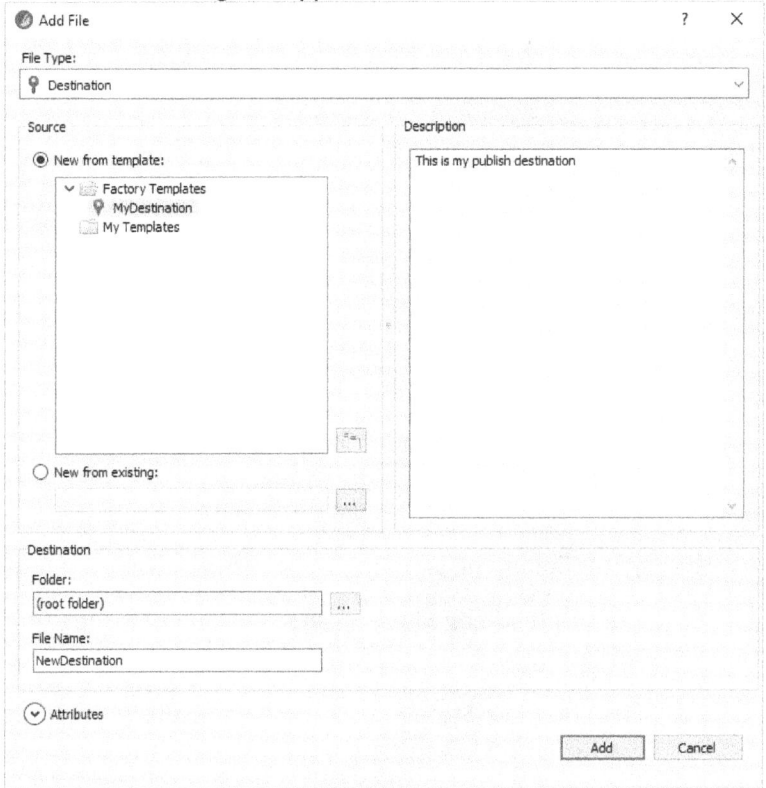

2 For **File Type**, select **Destination**.

3 Select a **Source** template.

4 Type a **File Name**.

5 Click **Add**.

The destination appears in the Destinations folder in the Project Organizer and opens in the Destination Editor.

6 For **Type**, select **File System**.

7 Click **Browse** to select a publishing directory.

8 Type a **Comment**.

9 Select **Upload Log File** to upload the publishing log file with your output files.

10 Select **Remove Stale Files** to remove deleted files from the destination when you republish.

11 Click **Save**.

To publish a target to a file server:

1 Open a target.

2 Select the **Publishing** tab.

3 Select one (or more) of the publishing destinations.

4 Click Publish.
Flare copies the generated files to the file server.

Publishing a target to SharePoint

You can publish any target to a SharePoint server.

Shortcut	Tool Strip	Ribbon
Ctrl+T	📝 (Content Explorer	File > New

To create a publishing destination for SharePoint:

1 Select **File** > **New**.
—OR—
Right-click the **Destinations** folder and select **Add Destination**.
The Add File dialog box appears.

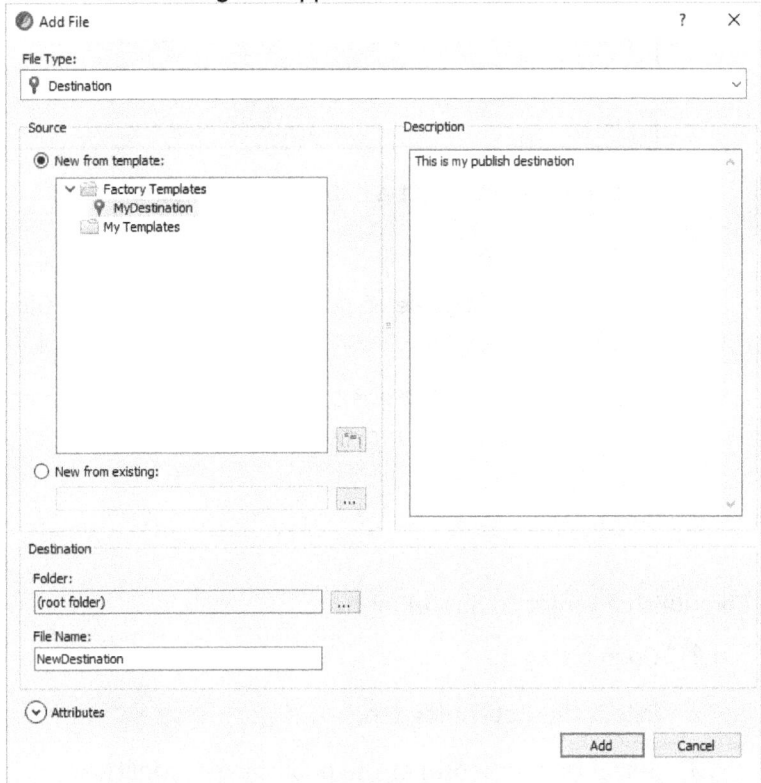

2 For **File Type**, select **Destination**.

3 Select a **Source** template.

4 Type a **File Name**.

5 Click **Add**.
The destination appears in the Destinations folder in the
Project Organizer and opens in the Destination Editor.

6 For **Type**, select **File System.**

7 Click **Browse** to select a publishing directory.

8 Type a **Comment.**

9 Select **Upload Log File** to upload the publishing log file with your output files.

10 Select **Remove Stale Files** to remove deleted files from the destination when you republish.

11 Click **Save.**

To publish a target to SharePoint:

1 Open a target.

2 Select the **Publishing** tab.

3 Select one (or more) of the publishing destinations.

4 Click 🕮 Publish.
Flare copies the generated files to the publishing destination.

Publishing a target to a web server

You can publish any target to a web server. HTML5, Clean XHTML, WebHelp, and PDF targets can be opened in a browser. Other targets will need to be downloaded.

Shortcut	Tool Strip	Ribbon
Ctrl+T	(Content Explorer	File > New

To create a publishing destination for a web server:

1 Select **File** > **New**.

   —OR—

   Right-click the **Destinations** folder and select **Add Destination**. The Add File dialog box appears.

   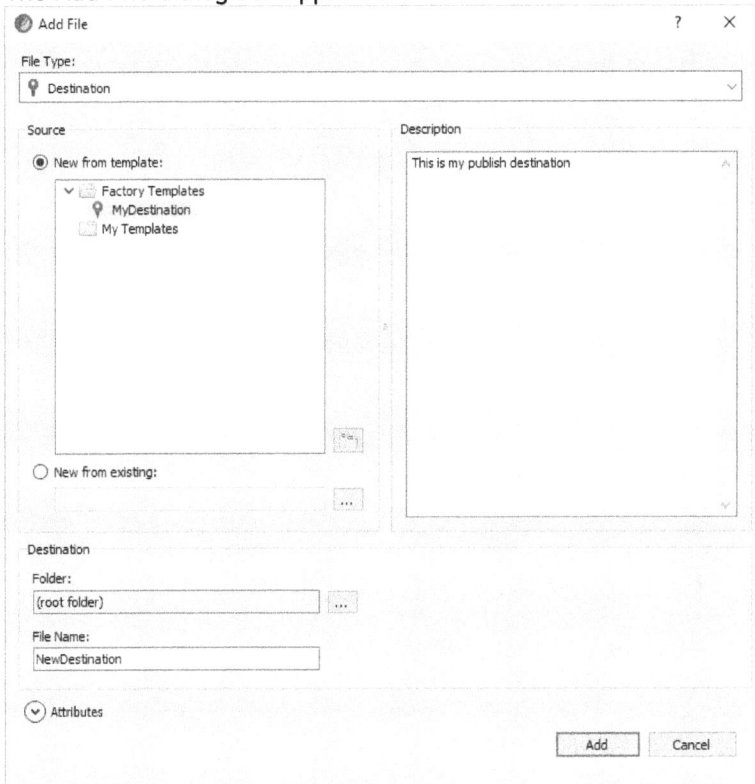

2 For **File Type**, select **Destination**.

3 Select a **Source** template.

4 Type a **File Name**.

5 Click **Add**.

The destination appears in the Destinations folder in the Project Organizer and opens in the Destination Editor.

6 For **Type**, select **FTP** or **SFTP**.

7 Type a **Host Name**.

8 If needed, type a **Port**.

9 Select a **Connection Mode**.

10 Type or select a **Directory**.

> 🗊 *Type "public_html" if you want to publish to the server's root directory.*

11 Click **Login Credentials** to provide your user name and password.

12 Select **Upload Log File** to upload the publishing log file with your output files.

13 Select **Remove Stale Files** to remove deleted files from the destination when you republish.

14 Click **Save**.

To publish a target to a web server:

1 Open a target.

2 Select the **Publishing** tab.

3 Select one (or more) of the publishing destinations.

4 Click 🌐 Publish.
Flare copies the generated files to the publishing destination.

Publishing a target to a source control server

You can publish any target to a source control server. For example, you could publish an HTML5 help system target to a source control server so that it can be included in a software application's build process.

Shortcut	Tool Strip	Ribbon
Ctrl+T	📑 (Content Explorer	File > New

To create a publishing destination for a source control server:

1 Select **File > New**.
—OR—
Right-click the **Destinations** folder and select **Add Destination**.

The Add File dialog box appears.

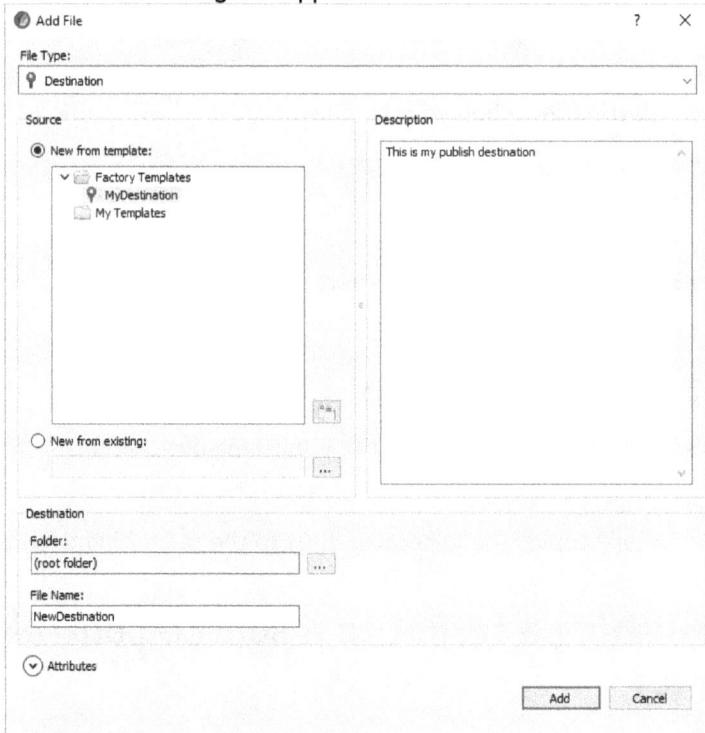

2 For **File Type**, select **Destination**.

3 Select a **Source** template.

4 Type a **File Name**.

5 Click **Add**.
   The destination appears in the Destinations folder in the
   Project Organizer and opens in the Destination Editor.

6 For **Type**, select **Source Control**.

7 Click **Browse** to select a publishing directory.

8 Type a **Comment**.

9 Click **Detect** to search for existing source control bindings. If Flare finds an existing binding for Git, Subversion, Perforce Helix Core, or TFS, it will automatically fill in the source control fields.

10 Click **Login Credentials** to provide your user name and password.

11 Select **Upload Log File** to upload the publishing log file with your output files.

12 Select **Remove Stale Files** to remove deleted files from the destination when you republish.

13 Click **Save**.

To publish a target to a source control server:

1 Open a target.

2 Select the **Publishing** tab.

3 Select one (or more) of the publishing destinations.

4 Click 🔘 Publish.
   Flare copies the generated files to the publishing destination.

Publishing a target to Central

If your project is bound to MadCap Central, you can publish your HTML5 target(s) directly to Central.

For information about binding a project to Central, see "Binding a project to Central" on page 442.

To publish a target to MadCap Central:

1 Open a target.

2 Select the **Publishing** tab.

3 Select the **Central** destination.

4 Click 🔘 Publish.
   Flare copies the generated files to Central.

◇ *The build will not be accessible by users unless you set it to Live in Central.*

Publishing a target to Salesforce

Because of limitations in Salesforce, you will need to build a Clean XHTML target to publish to Salesforce. You will also need to install MadCap Connect for Salesforce. You can install MadCap Connect for Salesforce by selecting the "Custom" installation option when you install Flare.

In Salesforce, you will need to create an article type that includes a rich text area field. For details about rich text area fields in Salesforce, see help.salesforce.com.

If you need to use a proxy server to connect to Salesforce, see "Connecting to a proxy server" on page 484.

Shortcut	Tool Strip	Ribbon
Ctrl+T	▦ (Content Explorer	File > New

To create a publishing destination:

1 Select **File** > **New**.

   —OR—

   Right-click the **Destinations** folder and select **Add Destination**.
   The Add File dialog box appears.

2 For **File Type**, select **Destination**.

3 Select a **Source** template.

4 Type a **File Name**.

5 Click **Add.**

The destination appears in the Destinations folder in the Project Organizer and opens in the Destination Editor.

6 For **Type,** select **MadCap Connect for Salesforce.**

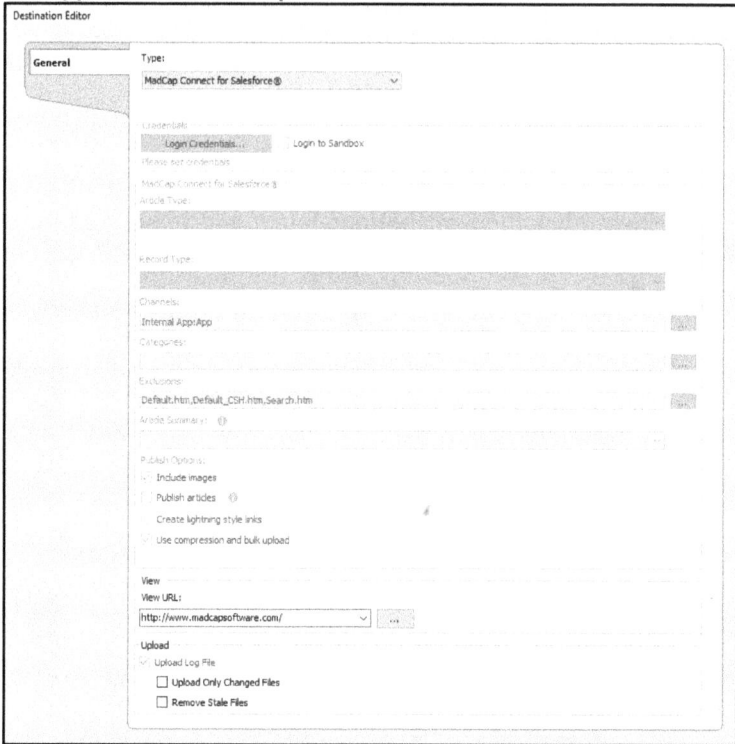

7 Click **Login Credentials** to provide your user name and password.

8 Select the **Article Type** that you created in Salesforce.

◇ *The article type must contain a rich text field.*

9 Select a **Rich Text Field.**

10 Select a **Record Type.**

11 Select the **Channel** that can view the content:
- □ Partner
- □ Customer
- □ Public Knowledge Base

12 If you have access to a **Shared Asset Library**, select the library so that Flare will store the publish log to the shared library. If you do not have access to a shared asset library, the publish log will be stored in your private library.

13 Type or select the **Categories** to use to classify your topics.

14 Specify any **Exclusions** to prevent topics from being published.

📖 *You can use asterisks to exclude multiple topics. For example, Sales*.htm would exclude all topics that start with the word "Sales."*

15 Type an **Article Summary** to be used for any topics that do not have a meta description. See "Adding meta descriptions" on page 212.

16 Select a **URL Naming** scheme.

17 Select **Include Images** to include images in your topics.

◇ *Each image must be less than 1MB, and Salesforce only supports the PNG, GIF, and JPEG image formats.*

18 Select **Publish Articles** to make your content viewable by the public.

19 If you are using Salesforce Lightning, select **Create Lightning Style Links**.

20 Select **Use Compression and Bulk Upload** to compress the output files and reduce the publishing time.

21 Select **Upload Log File** to upload the publishing log file with your output files.

22 Select **Remove Stale Files** to remove deleted files from the destination when you republish.

23 Click **Save**.

To publish a target to Salesforce:

1 Open a target.

2 Select the **Publishing** tab.

3 Select one (or more) of the publishing destinations.

4 Click 🖼 Publish.

Flare copies the generated files to the publishing destination.

Publishing a target to ServiceNow

Because of limitations in ServiceNow, you will need to build a Clean XHTML target to publish to ServiceNow. You will also need to install MadCap Connect for ServiceNow. You can install MadCap Connect for ServiceNow by selecting the "Custom" installation option when you install Flare.

Before you can publish to ServiceNow, a system admin must create a new entry for Flare in the ServiceNow application registry.

Shortcut	Tool Strip	Ribbon
Ctrl+T	🖳 (Content Explorer	File > New

To enable Flare publishing in ServiceNow:

1 In ServiceNow, select **System OAuth > Application Registry**.

2 Click **New**.

3 Select **Create an OAuth API Endpoint for External Clients**.

4 Type a name for the entry (e.g., "MadCap").

5 Select **Active**.

6 Write down the **Client ID** and **Secret**. You will need to provide this information when you setup the ServiceNow destination.

7 Click **Submit**.

To create a publishing destination:

1 Select **File** > **New**.

 —OR—

 Right-click the **Destinations** folder and select **Add Destination**. The Add File dialog box appears.

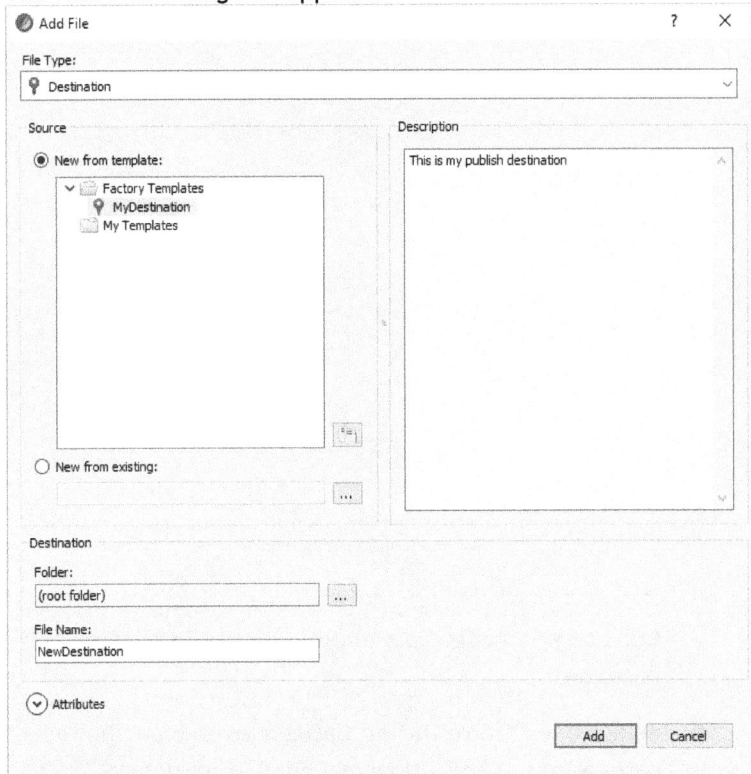

2 For **File Type**, select **Destination**.

3 Select a **Source** template.

4 Type a **File Name**.

5 Click **Add**.

 The destination appears in the Destinations folder in the Project Organizer and opens in the Destination Editor.

6 For **Type**, select **MadCap Connect for ServiceNow.**

7 Click **Login Credentials** to provide your user name and password.

8 Select **Use TOC to Define Categories** to base the resulting categories in ServiceNow on your TOC structure.

9 Select the **Default Knowledge Base** to store the published files.

10 Select the **Default Category** for the published files.

11 Specify any **Exclusions** to prevent topics from being published.

> 𝕋𝕀ℙ *You can use asterisks to exclude multiple topics. For example, Sales*.htm would exclude all topics that start with the word "Sales."*

12 Select a **Workflow** (Draft, Review, or Published) for the published files.

13 Select **Use Flare Keywords for ServiceNow Keywords** to convert Flare keywords to ServiceNow keywords.

◇ *Only Flare keywords without spaces can be added to ServiceNow.*

14 Select **Use Flare Concepts for ServiceNow Keywords** to convert Flare concepts to ServiceNow keywords.

◇ *Only Flare concepts without spaces can be added to ServiceNow.*

15 Select **Delete Stale ServiceNow Keywords** to remove previously published keywords that have been deleted in the topics.

16 Type the **Default Keywords** (separated by commas) that should be associated with all published articles.

17 Select **Use Flare Keywords for ServiceNow Tags** to convert Flare keywords to ServiceNow tags.

18 Select **Use Flare Concepts for ServiceNow Tags** to convert Flare concepts to ServiceNow tags.

19 Select **Delete Stale ServiceNow Tags** to remove previously published tags that have been deleted in the topics.

20 Type the **Default Tags** (separated by commas) that should be associated with all published articles.

21 Select **Upload Log File** to upload the publishing log file with your output files.

22 Select **Remove Stale Files** to remove deleted files from the destination when you republish.

23 Click **Save**.

To publish a target to ServiceNow:

1 Open a target.

2 Select the **Publishing** tab.

3 Select one (or more) of the publishing destinations.

4 Click ⊕ Publish.

Flare copies the generated files to the publishing destination.

Publishing a target to Zendesk

You can publish a Clean XHTML or "skinless" HTML5 target to Zendesk. You will also need to install MadCap Connect for Zendesk. You can install MadCap Connect for Zendesk by selecting the "Custom" installation option when you install Flare.

If you need to use a proxy server to connect to Zendesk, see "Connecting to a proxy server" on page 484.

Shortcut	Tool Strip	Ribbon
Ctrl+T	📄 (Content Explorer	File > New

To create a publishing destination for Zendesk:

1 Select **File > New**.

—OR—

Right-click the **Destinations** folder and select **Add Destination**.

The Add File dialog box appears.

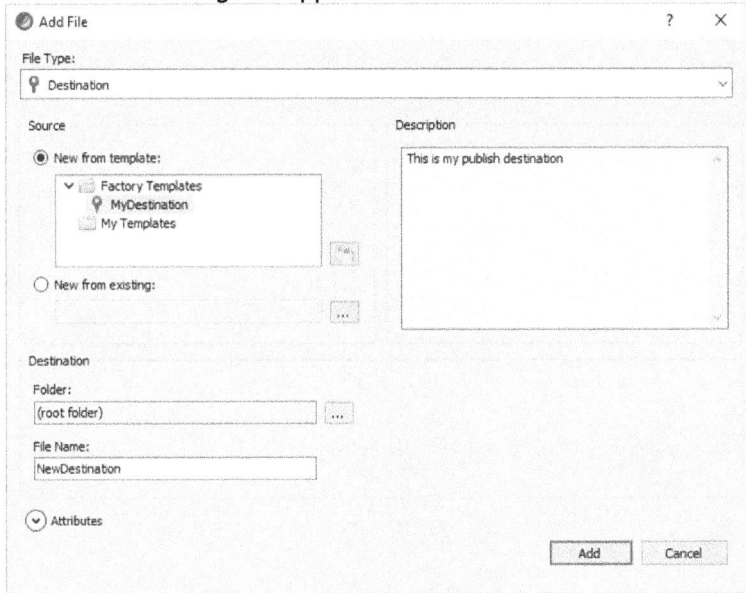

2 For **File Type**, select **Destination**.

3 Select a **Source** template.

4 Type a **File Name**.

5 Click **Add**.

The destination appears in the Destinations folder in the
Project Organizer and opens in the Destination Editor.

6 For **Type**, select **MadCap Connect for Zendesk**.

7 Click **Login Credentials** to provide your user name and password.

8 Select **Use TOC to Define** to structure your content by **Categories and Sections** or only by **Sections**.

9 Select whether you want to create sub-sections.

10 Select the **Default Category** you created in Zendesk.

11 Select the **Default Section** you created in Zendesk.

12 If needed, specify any **Exclusions** to prevent files from being published.

TIP *You can use asterisks to exclude multiple topics. For example, Sales*.htm would exclude all topics that start with the word "Sales."*

13 For **User Visibility**, select the user group who is allowed to view the content.

14 For **Management Permissions**, select the user group who is allowed to edit and publish content.

15 Select **Preserve Unsafe HTML** to include multimedia-related and other HTML tags that Zendesk considers unsafe. For more information about Zendesk's "unsafe" tags, see support.zendesk.com/hc/en-us/articles/115015895948-Allowing-unsafe-HTML-in-articles

16 Select **Publish Articles** to make your published content viewable in Zendesk. If this option is not selected, your content will be published in draft mode.

17 Select **Upload Log File** to upload the publishing log file with your output files.

18 Select **Delete Stale Files** to remove deleted files from the destination when you republish.

19 Click **Save**.

To publish a target to Zendesk:

1 Open a target.

2 Select the **Publishing** tab.

3 Select one (or more) of the publishing destinations.

4 Click Publish.
Flare copies the generated files to the publishing destination.

Batch generating targets

You can create a batch target to build and/or publish multiple targets. You can even schedule a batch target to run at a specific time every day, week, or month.

Shortcut	Tool Strip	Ribbon
Ctrl+T	📝 (Content Explorer	File > New

To create a batch target:

1 Select **File > New**.
  —OR—
  Right-click the **Targets** folder and select **Add Batch Target**.
  The Add File dialog box appears.

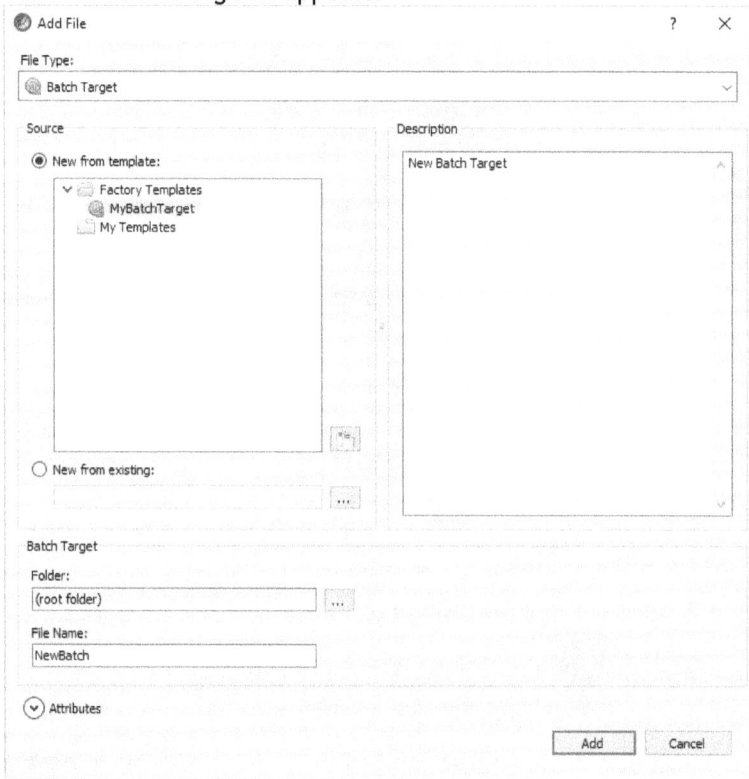

2 For **File Type,** select **Batch Target**.

3 Type a **File Name**.
  You don't have to type the .fltar extension. Flare will add it for
  you if you leave it out.

4 Click **Add.**

The batch target appears in the Targets folder in the Project Organizer and opens in the Target Editor.

5 On the **Targets** tab, select the target(s) you want to build and/or publish.

6 Click **Save.**

To schedule a batch target:

1 Open a batch target.

2 Select the **Schedule** tab.

3 Click **New.**

The New Trigger dialog box appears.

4 Select a frequency **Setting.**

5 Select a **Start** date and time.

6 If you selected a daily, weekly, or monthly frequency setting, select the recurrence details.

7 If the batch generate should repeat, select **Repeat task every** and specify how often and how long the repeating should occur.

8 If the repeating should expire, select **Expire** and specify an expiration date.

9 If you are ready to enable batch generation, select **Enable**.

10 Click **OK**.

Gathering usage data with Central

If you have a Central license, you can collect and view user analytics information for an HTML5 target in Central. Central provides the following reports:

- Browser statistics
- Context-sensitive help calls
- Operating system statistics
- Search phrases
- Search phrases with no results
- Viewed topics

To associate a Central analytics key with an HTML5 target:

1 Open an HTML5 target.

2 Select the **Analytics** tab.

3 For **Provider**, select **Central**.

4 If you are not logged in to Central, click **Login** and login.

5 For **Central License**, select your Central license.

6 Click **Create**, type a name, and click **OK**.
—OR—
Select a key.

Context-sensitive help

Context-sensitive help (or "CSH" if you like acronyms) is help that opens to a specific topic or micro content block based on where you are in an application. For example, a "Print Preview" dialog box could open a help topic about printing, and a "CVV" field could open a micro content block about CVV ("card verification code") numbers.

To create context-sensitive help, you need to associate your help topics and/or micro content blocks to the fields, windows, and/or dialog boxes that are used in your application. This process is called "mapping," and it uses two files: header files and alias files.

Adding a header file

A header file is used to assign a number and an identifier to each dialog box and window in an application. Many programming applications create header files automatically, but you can also create header files using Flare. Header files have a .h or .hh extension, and they appear in the Advanced folder in the Project Organizer.

Header files use the following format:
#define MyID *number*

For example:
#define Save_dialog 1000

Shortcut	Tool Strip	Ribbon
Ctrl+T	📄 (Content Explorer	File > New

To create a header file:

If your software team has created a header file, you can copy it to the Advanced folder in the Project folder. If you need to create a header file, follow these steps.

1 Select **File** > **New**.

 —OR—

 Right-click the **Advanced** folder and select **Add Header File**.
 The Add File dialog box appears.

2 For **File Type**, select **Header File**.

3 Select a **Source** template.

4 Type a **File Name** for the header file.
 You don't have to type the .h extension. Flare will add it for
 you if you leave it out.

5 Click **Add**.
 The header file appears in the Advanced folder in the Project
 Organizer and opens in the Text Editor.

Creating an alias file

Alias files are used to match identifiers to a help topic or micro content block. At first, an alias file might seem unnecessary: why not just put everything in the header file? The reason is that you need to share the header file with the development team. They need it to compile the application, and their programming application might automatically update the header file. By using an alias file, you can keep linking topics or micro content blocks to identifiers while the developers are creating the application.

In Flare, alias files can also be used to assign a skin to a help topic or micro content block when it is opened from the application. For example, you could normally open your help system in a large, 700x500 window with the navigation pane on the left. When you open it from a context-sensitive link, it could open in a smaller window without the navigation pane.

Alias files have a .flali extension. They appear in the Project Organizer in the Advanced folder.

Shortcut	Tool Strip	Ribbon
Ctrl+T	📄 (Content Explorer	File > New

To create an alias file:

1 Select **File** > **New**.
   —OR—
   Right-click the **Advanced** folder and select **Add Alias File**.
   The Add File dialog box appears.

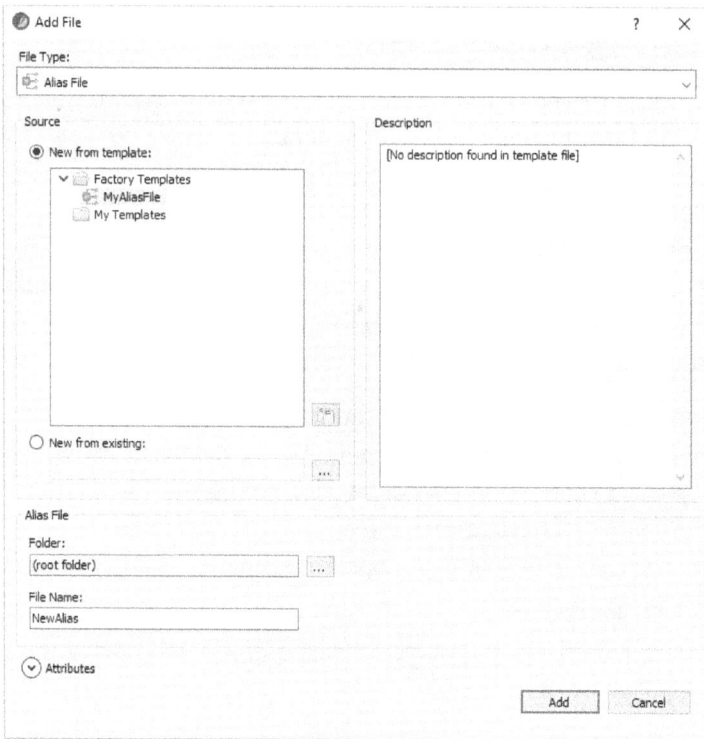

2 For **File Type**, select **Alias File**.

3 Select a **Source** template.

4 Type a **File Name** for the alias file.
You don't have to type the .flali extension. Flare will add it for you if you leave it out.

5 Click **Add**.
The alias file appears in the Advanced folder in the Project Organizer and opens in the Alias Editor.

To assign an identifier to a topic or micro content block:

1 Open an alias file.
Alias files are stored in the Advanced folder in the Project Organizer.

2 On the right side of the Alias Editor, select an identifier.

3   On the left side of the Alias Editor, select a topic or micro content block to link to the identifier.

4   Double-click the topic or micro content block.
    —OR—
    Click [icon].
    The selected topic's filename or micro content block's ID appears in the **File** column.

5   Continue assigning topics or micro content blocks to identifiers.

6   Save your alias file.
    You can open your alias file and add new identifiers at any time.

Testing context-sensitive help

After you assign identifiers to your topics, you can test your context-sensitive help links.

To test context-sensitive help:

1   Build your target.

2   Right-click your target and select **Test CSH API Calls**.
    The Context-sensitive Help API Tester dialog box appears.

3   Next to each identifier, click [Test].
    The correct help topic or micro content block should appear.

Finding topics without context-sensitive help IDs

You can use the Topics without Context-sensitive Help IDs report to find topics that do not have an assigned context-sensitive help ID.

To find topics without an assigned help ID:

☐   Select **Analysis > More Reports > Topics Not Linked by Map ID**.
    The Topics Not Linked by Map ID report appears.

Sample questions for this section

1 Which target types were created by MadCap software? (select all that apply)
   A) HTML5
   B) HTML Help
   C) WebHelp
   D) EPUB

2 What is Clean XHTML?
   A) An output format that does not include MadCap-specific tags, files, and scripts.
   B) A feature that removes inline formatting from your topics.
   C) An import option that cleans Word documents before they are imported.
   D) An imported XHTML file.

3 You can create a batch target to: (select all that apply)
   A) Build and publish HTML5 targets
   B) Build and publish PDF targets
   C) Combine HTML5 targets into one output
   D) Schedule automatic builds

4 What is a primary target?
   A) The main or "primary" target in merged projects.
   B) The only target you can create from a project.
   C) A project you plan to link to another project.
   D) The target you expect to build the most often.

5 How many targets can you set up in a project?
   A) One
   B) One for each type (i.e., one HTML5, one WebHelp, etc.)
   C) One for online and one for print.
   D) As many as you want.

6 What happens when you select **Project** > **Clean Project**?
   A) Flare fixes any incorrect XHTML code in your topics.
   B) Flare finds any files in the Content folder that are not being used.

C) Flare deletes everything in the Output folder.

D) Flare fixes any broken links.

7 What happens when you publish a target?

A) Flare creates the output files.

B) Flare copies the output files to a folder on a network or a web server.

C) Flare prints the output files.

D) Flare creates a postscript file that you can send to a printer.

8 What is the "Startup Topic?"

A) The topic that opens when you open the project in Flare.

B) The first topic that users see when they open the online target.

C) The file users double-click to open the target.

D) The file you double-click to open your project in Flare.

Project management

This section covers:

- [ ] Templates
- [ ] Spell check
- [ ] Thesaurus
- [ ] Text analysis
- [ ] Find and replace
- [ ] File tags
- [ ] Meta tags
- [ ] Reports
- [ ] Track changes
- [ ] Annotations
- [ ] Topics reviews
- [ ] Pulse
- [ ] Source control
- [ ] Central
- [ ] SharePoint integration
- [ ] Project archiving

Templates

Everything you create in Flare, including topics, stylesheets, template pages, and page layouts, is based on a template. Templates are just a "starting point" when you create a file. The new topic is not linked to the template, so any changes you make to the template will not be made to the topics that were created based on the template.

Creating a topic template

If your topics always have a similar structure, such as a heading, paragraph, table, and list, you can create a template to start your topics with these blocks of content.

Shortcut	Tool Strip	Ribbon
none	File > Save as Template	File > Save > Save as Template

To create a topic template:

1 Open the topic that you want to become a template.

2 Add any content you want to include in the template.

3 Format the content.

4 Select **File** > **Save** > **Save as Template**.
The Save as Template dialog box appears.

5 Type a **Template Name**.

6 Click **OK**.
Your template will be available in the My Templates folder. Topic templates use the same extension as topics: .htm.

TIP▷ *By default, templates are stored in the My Documents\My Templates\Content folder on your PC. If you work on a team, you can move your templates to a file server. When you create a new topic, select "New from existing" and click [...] to select the template on the file server.*

Creating a Contribution template

If other members of your team are using Contributor to create topics, you can create a Contribution template. A Contribution template can include pre-formatted text, images, and tables to help them create topics using the correct styles.

Shortcut	Tool Strip	Ribbon
none	Tools > Contributions > Create Contribution Template	File > Save > Save as Contribution Template

To create a Contribution template:

1 Open the topic that you want to become a template.

2 Add any content to the topic that you want to include.

3 If you want to prevent a user from modifying part of the topic:

    ☐ Highlight the content.

    ☐ Select **Review** > **Lock**.

4 Select **File** > **Save** > **Save as Contribution Template**.
The Save Topic as Contribution Template wizard appears.

5 Type a **Template Name**.

6 Select a template location.
By default, Contribution templates are stored in the Documents\My Templates\Content folder on your PC.

7 Click **Next**.

8 Select any content files the template uses, such as images, page layouts, template pages, snippets, or stylesheets.

9 Click **Next**.

10 Select any condition tag sets or variable sets the template uses.

11 Click **Finish**.

The Contribution template is saved. Contribution templates use the .mccot extension.

12 Click **Yes** if you want to email the Contribution template to a contributor.

TIP> *If you work on a team, you can move your Contribution templates to a file server. When contributors create a new topic, they can select "New from existing" and click ⊡ to select the Contribution template on the file server.*

Using a project template

MadCap provides numerous project templates to help you get started creating different types of projects, including policies and procedures, knowledge bases, user guides, help systems, and brochures. They include sample content and basic designs you can use "as is" or modify as needed. MadCap's project templates provide professionally designed HTML5 targets that you can use for new projects or apply to existing projects.

You can download the project templates at www.madcapsoftware.com/downloads/madcap-flare-project-templates.

Spell check, thesaurus, and text analysis

You can spell check a topic, the entire project, or while you are editing a topic in the XML Editor. You can also create a list of terms to ignore when spell checking and add language dictionaries.

Spell checking a topic or project

Shortcut	Tool Strip	Ribbon
F7	Tools > Spell Check Window	Tools > Spell Check Window

To spell check a topic or project:

1   Open a topic.

2   Select **Tools > Spell Check Window**.
    —OR—
    Press **F7**.
    The Spell Check window appears.

3   For **Spell Check**, select a location and type of file to spell check.

4   Click **Start**.

5   For each potential misspelled word, click **Ignore**, **Ignore All**, **Add to Dictionary**, **Change**, or **Change All**.

6   If you select Add to Dictionary, select whether you want to add the word to a global or project-specific dictionary.

To enable or disable spell check while typing:

☐   Select **Tools > Spell Check While Typing**.

Ignoring words when spell checking

You can create a list of terms that you want to ignore when spell checking. For example, you might ignore acronyms, words with numbers, or technical and jargon terms.

Shortcut	Tool Strip	Ribbon
Alt+T, I	Tools > Ignored Words	Tools > Ignored Words

To ignore words when spell checking:

1 Select **Tools** > **Ignored Words**.
The Ignored Words dialog box appears.

2 Type the words you want to ignore.

3 Click **OK**.

Setting spell check options

You can set the spell checker to ignore certain types of words, like uppercase words, words that use a specified style, or words that have been tagged with a specified condition.

Shortcut	Tool Strip	Ribbon
Alt+F, T	Tools > Options	File > Options

To ignore words when spell checking:

1 Select **File** > **Options**.
The Options dialog box appears.

2 Select the **Spelling** tab.

3 In the **Spelling Options** section, select the types of words you want to ignore.

4 If you want to ignore content that is formatted with a style, click inside the **Ignored styles** text box and type the style's name.

5 If you want to ignore content that is tagged with a condition, click inside the **Ignored conditions** text box and type the condition tag's name.

6 Click **OK**.

Adding spell check dictionaries

Flare includes dictionaries for numerous languages and country-specific language variants. You can download and install open-source dictionaries for other languages.

Many dictionaries are available at: extensions.services.openoffice.org/dictionary

Shortcut	Tool Strip	Ribbon
Alt+F, T	Tools > Options	File > Options

To add a spell checking dictionary:

1 Select **File** > **Options**.
The Options dialog box appears.

2 Select the **Spelling** tab.
Installed dictionaries are indicated with a green dot in the Spell column.

3 Click **Import Dictionaries**.
The Select dictionary files dialog box appears.

4 Select a dictionary.
Dictionary files have a .oxt extension.

5 Click **Open**.

Sharing a global dictionary

Flare supports project-specific and global dictionaries. Project dictionaries are stored in the Project Organizer in the Advanced folder. Global dictionaries are stored in the AppData folder in Windows by default. However, you can select a different location for the global dictionary.

Shortcut	Tool Strip	Ribbon
Alt+F, T	Tools > Options	File > Options

To share a global dictionary:

1 Select **File > Options**.
   The Options dialog box appears.

2 Select the **Spelling** tab.
   Installed dictionaries are indicated with a green dot in the Spell column.

3 For **Select Global Dictionary Location**, select **Use Custom Location**.

4 Click browse, select a folder, and click **Select Folder**.

5 Click **Open**.

Reviewing readability and grade level scores

You can view readability scores, grade level scores, and other detailed readability information for a topic or score averages for a selected group of topics.

Shortcut	Tool Strip	Ribbon
Alt+T, R	Tools > Text Analysis	Tools > Text Analysis

To review readability and grade level scores:

1 Select **Tools > Text Analysis**.
   The Text Analysis window appears.

2 For **Analysis Options**, select the information you want to review.

3 Select an **Analyze Text In** option.

4 Click **Analyze.**

Finding synonyms

You can use Flare's thesaurus to find potential synonyms for terms in your topics.

Shortcut	Tool Strip	Ribbon
Alt+T, T, H	Tools > Thesaurus	Tools > Thesaurus

To find synonyms for a term:

1 Select **Tools** > **Thesaurus.**
The Thesaurus window appears.

2 Type a term.

3 Click **Search.**

Find and replace

You can find or find and replace content within a topic, all of the topics in a folder, or all of the topics in a project. Flare includes standard options such as case sensitivity and searching in the source code, it also provides advanced options such as using wildcards and regular expressions. You can also find elements, such as HTML tags or Flare-specific features such as drop-downs.

Finding content in a topic

Shortcut	Tool Strip	Ribbon
Ctrl+F	Edit > Find and Replace	Home > Quick Find

To find content in a topic:

1 Select **Home** > **Quick Find**.
   —OR—
   Press **Ctrl+F**.
   The Quick Find widget appears.

2 In the text box, type the text you want to find.

3 If you want to only match whole words, click and select **Whole word**.

4 If you want to make the find case sensitive, click and select **Match case**.

5 If you want to use wildcard or regular expressions, click and select **Wildcards** or **Regular expressions**.

6 Click **Find Previous** or **Find Next**.

Finding and replacing in a topic

Shortcut	Tool Strip	Ribbon
Ctrl+H	Edit > Find and Replace	Home > Quick Replace

To find and replace content in a topic:

1 Select **Home** > **Quick Replace**.
   —OR—
   Press **Ctrl+H**.
   The Quick Replace widget appears.

2 In the **Find** text box, type the text you want to find.

3 In the **Replace** text box, type the text you want to add.

4 If you want to only match whole words, click and select **Whole word**.
   —OR—

   If you want to make the find case sensitive, click and select **Match case**.
   —OR—

   If you want to use wildcard or regular expressions, click and select **Wildcards** or **Regular expressions**.

5 Click **Replace Next** or **Replace All**.

Finding and replacing in multiple topics

Shortcut	Tool Strip	Ribbon
Ctrl+Shift+F	Edit > Find and Replace	Home > Find and Replace Text

To find and replace content in multiple topics:

1 Select **Home** > **Find and Replace Text**.
   —OR—

Press **Ctrl+Shift+F**.
The File and Replace in Files window appears.

2   In the **Find** text box, type the text you want to find.

3   In the **Replace with** text box, type the text you want to add.

4   Select a **Find in** option.

5   Select the type of file(s) you want to search.

6   If the options are closed, click **Options**.

7   Select **Match case** if you want to make the find case sensitive.

8   Select **Whole word** if you want to only match whole words and not parts of words.

9   Select **Find in source code** if you want to also find matches in the code.

10   For **Search type**, select **Wildcards** or **Regular expressions** to create an advanced find expression.

11   Select results window 1 or 2.

12   Click **Replace** or **Replace All**.
The results appear in the results window.

Using wildcards and regular expressions

You can use wildcards and regular expressions to create advanced find expressions. Wildcards are useful for partial searches, such as finding all phone numbers with a specified area code. Regular expressions are much more advanced and can be used to find all h1s or spans or a word if it is near another word.

Wildcard examples

Find expression	Result
404-*	Finds "404-555-5555," "404-555-0000", etc.
.*.com	Finds "mysite.com," "yoursite.com," etc.

Find expression	Result
locali?e	Finds "localize" and "localise"
locali*	Finds "localize," "localization," "localise," and "localisation"

For more information about wildcards, see en.wikipedia.org/wiki/Wildcard_character

Regular expression examples

Find expression	Result
<h1\b[^>]*>(.*?)</h1>	Finds all h1s
\bquick\W+(?:\w+\W+){1,5}?fox \b	Finds the word "quick" only if it is within five words of the word "fox"

Flare supports .NET Framework regular expressions. For more information, see msdn.microsoft.com/en-us/library/hs600312.aspx

Finding and replacing elements

In addition to finding content, you can use the Find Elements window to find inline styles, HTML tags, or MadCap-specific tags.

Shortcut	Tool Strip	Ribbon
Ctrl+N	Edit > Find Elements	Home > Find Elements

To find elements:

1 Select **Home > Find and Replace Elements**.
—OR—
Press **Ctrl+N**.
The File Elements window appears.

2 For **Find**, select the type of content that you want to find.

3 For **Replace/Action**, select the type of content you want to add.

4 Select results window 1 or 2.

5 Click **Replace** or **Replace All.**

The results appear in the results window.

File tags

You can create file tags to specify authors, status levels, or any other type of information for files in your project. For example, you could create a "reviewer" tag to track who the subject-matter expert reviewer should be for each topic. File tags can be assigned to any type of file in your project. Unlike meta tags, file tags are not included in targets.

You can sort files by their assigned file tag(s) in the File List (View> File List). You can also create an information report of used and/or unused file tags.

Creating a file tag set

Flare provides templates to create author and status file tags, and you can modify them as needed. Or, you can create your own file tag set.

Shortcut	Tool Strip	Ribbon
Ctrl+T	📄 (Content Explorer	File > New

To create a file tag set:

1   Select **File** > **New**.
    The Add File dialog box appears.

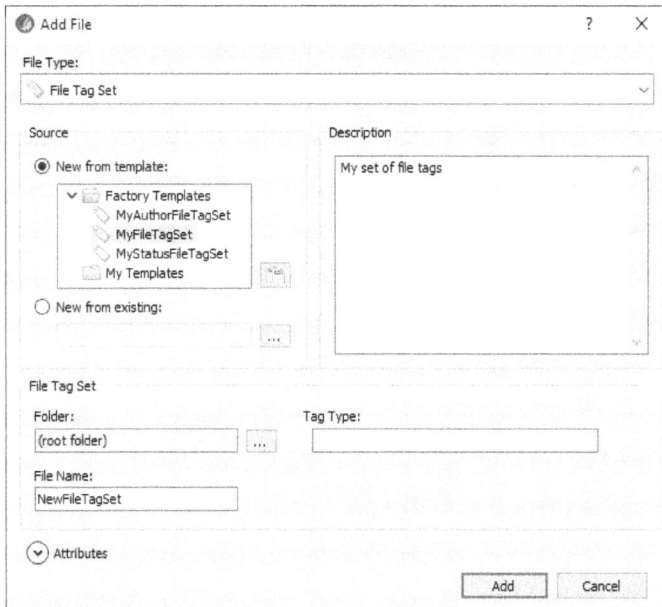

2 For **File Type**, select **File Tag Set**.

3 Select a **Source** template.

4 Type a **File Name**.

5 Click **Add**.
The file tag set file appears in the Advanced folder in the Project Organizer and opens in the File Tag Set Editor.

Creating a file tag

You can create file tags to track information about your files, such as status levels, that you don't need to include in the output.

To create a fie tag:

1 Open a file tag set.

2 Click 📄.
The new tag appears.

3 Type a name for the tag.

Setting a file tag's value in a file

You can set a value for a file tag in a file or for a folder. If you set a value for a folder, all of the files in the folder will inherit the value.

To set a file tag's value in a file or folder:

1 Right-click a file or folder and select **Properties**.
The Properties dialog box appears.

2 Select the **File Tags** tab.

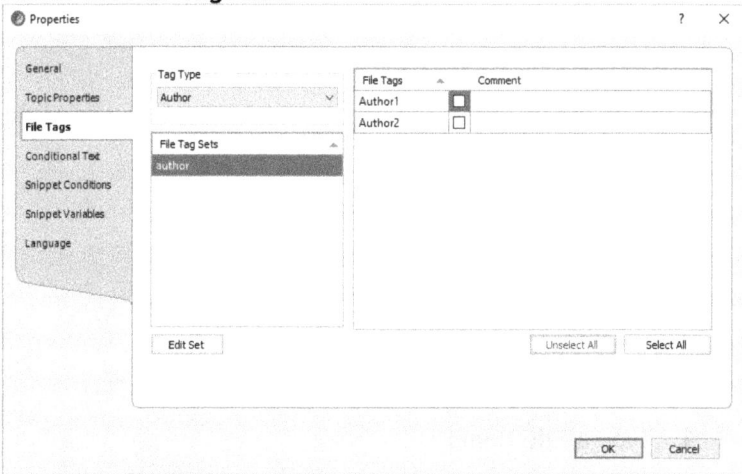

3 Select a **Tag Type**.

4 Select a **File Tag**.

5 Click **OK**.

To set a file tag's value when you create a file:

1 Select **File** > **New**.
The Add File dialog box appears.

2 Select a **File Type**, **Template**, and **Folder**.

3 Type a **File Name**.

4 Click **Attributes**.

5   For **File Tags**, click the ... (browse) link.
    The File Tags dialog box appears.

6   Select a **Tag Type**.

7   Select a **File Tag Set**.

8   Select a **File Tag** (or tags) and click **OK**.

Reviewing how file tags are used in a project

You can open the Used File Tags report to review how file tags are used in a project.

To review how file tags are applied in a project:

☐   Select **Analysis** > **Used Items** > **Used File Tags**.
    The Used File Tags report appears.

Meta tags

You can create meta tags to include "hidden" information inside the head section of your HTML5 your topics. The most commonly used meta tag is "description." If a topic contains a value for the description meta tag, the value appears in the search results. If a topic does not contain a description value, the first ~155 characters are used for the description in the search results.

You can create text or list meta tags. A text meta tag allows you to type a short value. A list meta tag allows you to select a value from a pre-defined set of options. If you already use file tags and want to include your file tags in HTML5 output, you can link a meta tag to a file tag set.

Creating a meta tag set

You can create multiple meta tag sets to organize meta tags and/or to use them to control micro content results.

Shortcut	Tool Strip	Ribbon
Ctrl+T	📑 (Content Explorer	File > New

To create a meta tag set:

1 Select **File** > **New**.
The Add File dialog box appears.

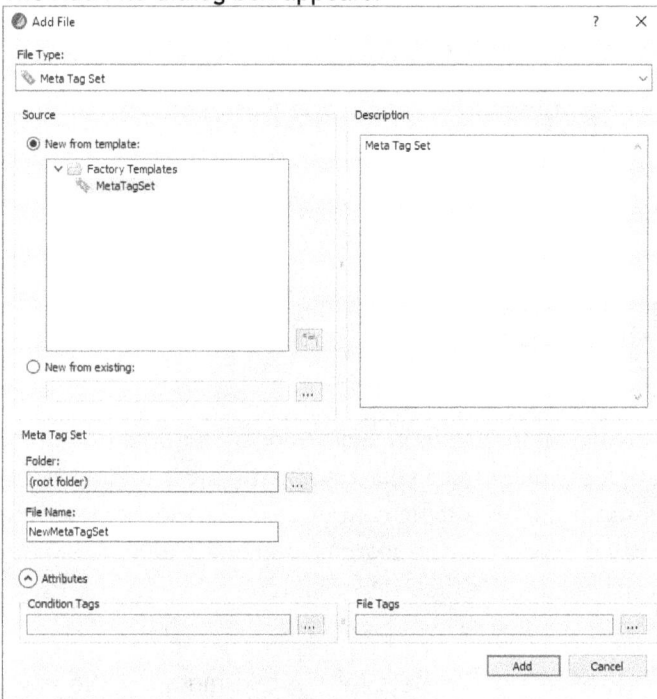

2 For **File Type**, select **Meta Tag Set**.

3 Select a **Source** template.

4 Type a **File Name**.

5 Click **Add**.

The meta tag set file appears in the Advanced folder in the Project Organizer and opens in the Meta Tag Set Editor.

Creating a meta tag

If a meta tag can have any value, like a description, you can create a text meta tag. If a meta tag's value should be limited to certain values, you can create a list meta tag.

To create a meta tag:

1 Open a meta tag set.

2 Click 📄.

3 Select **Text** or **List**.
A new meta tag is created.

4 Type a **Name** for the meta tag.

5 If you are creating a list meta tag, type a value for the list item.

6 Click ➕ to add additional list items.

7 If needed, type a default **Value** (text meta tag) or select a default list item (list meta tag).

Linking a meta tag to a file tag

You can link a meta tag to a file tag set to reuse file tag information and include it in targets. If you modify the file tag set, the linked meta tag will be automatically updated.

To link a meta tag to a file tag set:

1 Open a meta tag set.

2 Click 📄.

3 Select **File Tags** > and the name of the file tag set.
A new meta tag is created, and it is linked to the file tag set.

4 Type a **Name** for the meta tag.

Setting a meta tag's value in a topic

You can set a value for a meta tag in a file or for a folder. If you set a value for a folder, all of the files in the folder will inherit the value.

To set a meta tag's value in a file or folder:

1   Right-click a file or folder and select **Properties**.
    The Properties dialog box appears.

2   Select the **Meta Tags** tab.

3   Type a **Value** for the meta tag.

    TIP *Click* ⊠ *to use a variable in a meta tag's value.*

Overriding a meta tag's value in a target

You can specify a meta tag's definition in a target to override the default value set in the meta tag set.

To override a meta tag's definition in a target:

1   Open a target.

2   Select the **Meta Tags** tab.

3   Select the **Override values set in content files** checkbox for the meta tag.

4   Type a **Value**.

Reviewing how meta tags are used in a project

You can open the Used Meta Tags report to review how meta tags are used in a project.

To review how meta tags are applied in a project:

☐   Select **Analysis > Used Items > Used Meta Tags**.
    The Used Meta Tags report appears.

Reports

You can create analysis and information reports to manage and troubleshoot projects. Analysis reports are more powerful than information reports. For example, the Broken Links analysis report can be used to find and fix broken links. The Broken Links information report only identifies broken links. However, you can customize how information reports are formatted and specify the information that is included. You can save the data from either type of report, but information report data can also be archived in Flare.

Analysis reports

The analysis reports include:

Summary

Links

☐ Broken Links

☐ Broken Bookmarks

☐ External Links

☐ Absolute Links

☐ Named Destinations

☐ Index Keyword Links

☐ Concept Links

Suggestions

☐ Index Keyword Suggestions

☐ Snippet Suggestions

☐ Variable Suggestions

☐ New Style Suggestions

☐ Replace Local Style Suggestions

☐ Accessibility Suggestions

☐ Markup Suggestions

☐ Cross Reference Suggestions

☐ Writing Suggestions

☐ Frequent Segments

☐ Similar Segments

Undefined Items

☐ Undefined Glossary Term Links

☐ Undefined Variables

☐ Undefined Condition Tags

☐ Undefined File Tags

☐ Undefined Styles

☐ Undefined Meta Tags

Used Items

☐ Used Condition Tags

☐ Used File Tags

☐ Used Language Tags

- Used Variables
- Used index Keywords
- Used Concepts
- Used Bookmarks
- Used Meta Tags

More Reports

- Files with Annotations
- Files with Changes
- Topics Not in Index
- Topics Not in Selected TOC
- Topics Not Linked by Map ID

- Duplicate TOC Items
- Duplicate Style Formats
- Duplicate Meta Tags
- Unused Items
- Non-XML Topics
- Files With Snippet Conditions
- Files Without File Tags
- Topics Without Concepts
- Invalid Meta Values

File Issues

To open an analysis report:

1 Select the report from the Analysis menu.

2 If the report's data does not appear:

- Select **File > Options**.
- Select the **Project Analysis** tab.
- Select the **Advanced Scan Options** checkboxes.
- Click **OK**.

3 To save the report as a CSV file, click 🗔.

Information reports

Information reports include:

Bookmarks

- Broken bookmarks
- Unused bookmarks
- Used bookmarks

Concepts

- Concept links
- Files with concepts
- Topics with concept links

- ☐ Topics with concept links missing a concept
- ☐ Topics without concepts
- ☐ Used concepts
- ☐ Used search filters

Condition tags

- ☐ Applied conditions
- ☐ Files with condition tags
- ☐ Topics with snippet conditions
- ☐ Undefined condition tags
- ☐ Unused condition tags
- ☐ Used condition tags

Content files

- ☐ Empty content folders
- ☐ File word count
- ☐ Files with annotations
- ☐ Files with changes (includes annotations)
- ☐ Files with concepts
- ☐ Files with condition tags
- ☐ Files with equations
- ☐ Files with file tags
- ☐ Files with glossary term links
- ☐ Files with images
- ☐ Files with keywords
- ☐ Files with micro content links
- ☐ Files with language tags
- ☐ Files with multimedia
- ☐ Files with named destinations
- ☐ Files with QR codes
- ☐ Files with snippets

- ☐ Files with variables
- ☐ Files without file tags
- ☐ Topics without concepts
- ☐ Unused content files

Context-sensitive help

- ☐ Assigned CSH IDs
- ☐ Duplicate Map IDs
- ☐ Topics Linked By Map ID
- ☐ Topics Not Linked By Map ID
- ☐ Unused CSH IDs

File tags

- ☐ Files with file tags
- ☐ Files without file tags
- ☐ Undefined file tags
- ☐ Unused file tags
- ☐ Used file tags

Glossary term links

- ☐ Files with glossary term links
- ☐ Undefined glossary term links
- ☐ Used glossary term links

Images

- ☐ Files with images
- ☐ Unused images
- ☐ Used images

Index

- ☐ Files with keywords
- ☐ Index keyword links
- ☐ Index keyword suggestions

- Topics not in index
- Topics with index keyword links
- Topics with keyword links missing a keyword
- Used index keywords

Language tags

- Files with language tags
- Used language tags

Links

- Absolute links
- Broken bookmarks
- Broken links
- Broken snippet links
- Concept links
- Cross reference suggestions
- External links
- Index keyword links
- Named destinations
- Topics linked by map ID
- Topics not linked
- Topics not linked by map ID
- Topics with concept links
- Topics with concept links missing a concept
- Topics with index keyword links
- Topics with keyword links missing a keyword
- Undefined glossary term links
- Unused bookmarks
- Used bookmarks

- Used glossary term links

Meta tags

- Duplicate meta tags
- Invalid meta values
- Undefined meta tags
- Unused meta tags
- Used meta tags

Micro content

- Duplicate micro content phrases
- Files with micro content links
- Micro content empty phrases
- Micro content empty responses

Multimedia

- Files with multimedia
- Unused multimedia
- Used multimedia

Project

- Database errors
- Files with annotations
- Statistics

Snippets

- Broken snippet links
- Files with snippet conditions
- Files with snippets
- Snippet suggestions
- Topics with snippet conditions
- Unused snippets
- Used snippets

Styles

- ☐ Duplicate styles
- ☐ New style suggestions
- ☐ Replace local style suggestions
- ☐ Undefined styles
- ☐ Unused styles
- ☐ Used stylesheets

Target

- ☐ Condition tag not set
- ☐ Variables not overridden
- ☐ Word count

TOC

- ☐ Duplicate TOC items
- ☐ TOC - primary target
- ☐ Topics not in any TOC

Topics

- ☐ Accessibility suggestions
- ☐ Broken bookmarks
- ☐ Broken links
- ☐ Cross reference suggestions
- ☐ File word count

- ☐ Files with snippet conditions
- ☐ Markup suggestions
- ☐ Non-XML topics
- ☐ Topics linked by map ID
- ☐ Topics not in any TOC
- ☐ Topics not in index
- ☐ Topics not linked
- ☐ Topics not linked by map ID
- ☐ Topics with concept links
- ☐ Topics with concept links missing a concept
- ☐ Topics with keyword links
- ☐ Topics with keyword links missing a keyword
- ☐ Topics with meta tags
- ☐ Writing suggestions

Variables

- ☐ Files with variables
- ☐ Topics with snippet variables
- ☐ Undefined variables
- ☐ Unused variables
- ☐ Used variables
- ☐ Variable suggestions

To create an information report:

1 Select **File** > **New**.
   —OR—
   Right-click the **Reports** folder and select **Add Report File**.

The Add File dialog box appears.

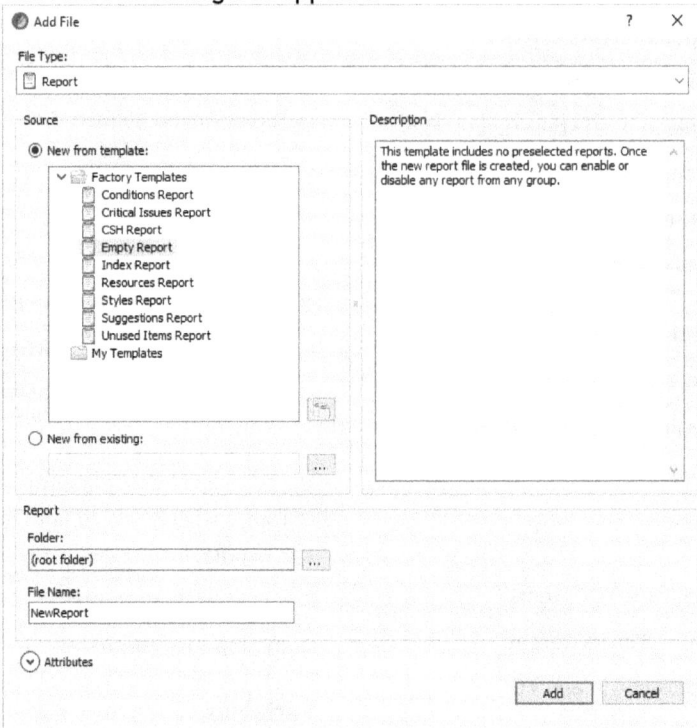

2   For **File Type**, select **Report File**.

3   Select a **Source** template.

4   Type a **File Name**.

5   Click **Add**.
    The report appears in the Reports folder in the Project
    Organizer and opens in the Report Editor.

To generate, archive, or print an information report:

1   Open an information report.

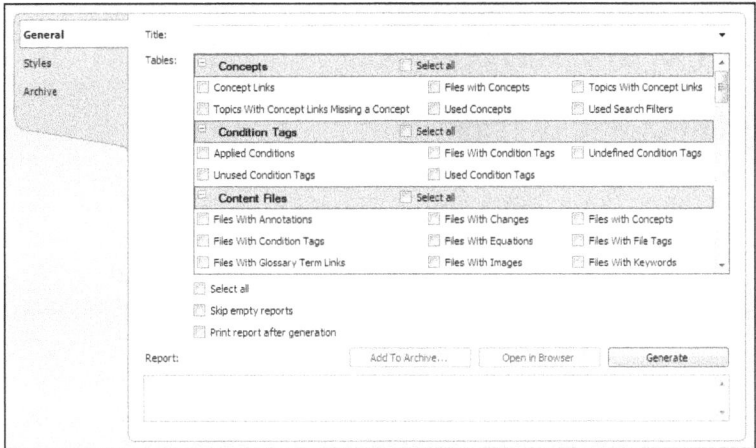

2 On the **General** tab, type a **Title** for the report.

3 Select the **Tables** to include in the report.

4 If you want to change the report's formatting, click the **Styles** tab and change the style properties.

5 On the **General** tab, click **Generate**.

6 To archive the report's information, click **Add to Archive**.

7 To print or save the report, click **Open in Browser**.

Track changes, annotations, and reviews

Tracking changes

You can enable track changes to capture changes made to a topic, snippet, or template page. Track changes are very useful during topic reviews, since the reviewers can see exactly how the content has changed. They will automatically appear in Contributor, and you can include them when you build a PDF or Word target.

✎ *When you enable track changes, it's enabled for all topics, snippets, and template pages. In older versions of Flare, enabling track changes only applied to the current file and any files you opened after enabling track changes.*

Shortcut	Tool Strip	Ribbon
Ctrl+Shift+E	📝 (Review toolbar)	Review > Track Changes

To track changes:

1   Open a topic.

2   Select **Review** > **Track Changes**.

3   Change the topic as needed.

To hide or show tracked changes:

1   Open a topic.

2   Select **Review** > **Show Changes**.

To accept or reject a tracked change:

1   Open a topic.

2 Select **Review** > **Show Changes**.

3 To select a specific change, click the change and select **Review** >
Accept Change.
To accept all changes, select **Review** > **Accept All Changes**.

To customize how changes display:

1 Select **File** > **Options**.
The Options dialog box appears.

2 Select the **Review** tab.

3 Type your **User Name**.

4 Type your **Initials**.

5 Select a **Tracked Changes Display** option.

6 If you want to include change bars, set the **Change Bar** option
to **On**.

7 Select a **User Colors** option.

To find files with tracked changes:

☐ Select **Analysis** > **More Reports** > **File with Changes**.
The Files with Changes report appears.

To include tracked changes in a PDF or Word target:

1 Open a PDF or Word target.

2 Select the **Advanced** tab.

3 Select **Preserve tracked changes**.

4 To select a specific change, click the change and select **Review** >
Accept Change.
To accept all changes, select **Review** > **Accept All Changes**.

Adding an annotation

You can add annotations to your topics to keep notes or to track your development progress. If you send topics for review, reviewers can use Contributor to add annotations. Annotations are not included when you build a target.

Shortcut	Tool Strip	Ribbon
Alt+R, W	(Review toolbar)	Review > Insert Annotation

To add an annotation:

1 Open a topic.

2 Select the content that you would like to annotate.

3 Select **Review** > **Insert Annotation**.
The Edit Annotation Pane appears.

4 If this is your first annotation, type your initials.

5 Type your annotation.

To show or hide annotations:

1 Open a topic.

2 Select **Review** > **Show Changes**.

To find files with annotations:

☐ Select **Analysis** > **More Reports** > **Files with Annotations**.
The Files with Changes report appears.

Locking content

You can lock content so that it cannot be edited. For example, you might lock a paragraph so that a reviewer using Contributor can see it but not change it.

To lock content so that it cannot be edited:

1 Open a topic.

2 Position your cursor inside the content you want to lock.

3 Select **Review > Lock**.
The paragraphs are locked and cannot be edited.

Sending topics for review in Contributor

You can create a review package to send one topic/snippet or multiple topics and/or snippets to reviewers who use Contributor. If your topics contain snippets, the snippets will be automatically included in the review package.

If you send your topics using the TOC, Flare will include the TOC and all of the topics that are linked to by the selected book and its sub-books and pages. Reviewers can use Contributor to revise your topics, add annotations, and return the reviewed topics.

Shortcut	Tool Strip	Ribbon
Alt+R, S	▤◿ (Review toolbar)	Review > Send for Review

To send topics for review in Contributor:

1 Select **Review > Send for Review**.
The Send Files for Review wizard appears.

2 Type a **Review Package Name**.

3 Type a **Review Package Description**.

4 Click ⊕.
The Open File dialog box appears.

5 Select the topics you want to send for review and click **Open**.

◇ *If a selected topic contains a snippet, the snippet will also be added to the review package.*

6 Select a **TOC** to include with your review package.

7 Click **Next**.

8 Select **Send to MadCap Contributor**.

9 If you have defined multiple definitions for your variables and selected different definitions for your targets, select a target to specify the variable definitions that will appear in your review topics.

10 If you use condition tags in the selected topics, click to include or exclude content in your review topics based on their condition tags.

11 If you want to save the review package and send it to the reviewer(s) yourself, click **Save**.

If you want to email the review package:

    ☐ Click **Next**.

    ☐ Type a subject and message for the email.

    ☐ Click **Add Email Recipient** to select a recipient.

    ☐ Click **Send**.
      Your email application opens.

    ☐ Send the email.

To send topics for review using the TOC:

1 Open a TOC.

2 Right-click a book and select **Send for Review**.
The Send Files for Review wizard appears.

3 Type a **Review Package Name**.

4 Type a **Review Package Description**.

5 Click **Next**.

6 Select **Send to MadCap Contributor**.

7   If you have defined multiple definitions for your variables and selected different definitions for your targets, select a target to specify the variable definitions that will appear in your review topics.

8   If you use condition tags in the selected topics, click to include or exclude content in your review topics based on their condition tags.

9   If you want to save the review package and send it to the reviewer(s) yourself, click **Save**.

If you want to email the review package:

- □   Click **Next**.

- □   Type a subject and message for the email.

- □   Click **Add Email Recipient** to select a recipient.

- □   Click **Send**.
  Your email application opens.

- □   Send the email.

Accepting review edits from Contributor

When the review is finished, you can accept or reject the review comments and accept the reviewed topic/snippet. When you accept the reviewed topic or snippet, the original file is replaced by the reviewed file.

Shortcut	Tool Strip	Ribbon
Alt+R, T	(Review toolbar)	Review > Topic Reviews

To accept a reviewed topic or snippet from Contributor:

1   Select **Review** > **File Reviews**.
The File Reviews window appears.

2 In the text box at the top of the window, select **Inbox**.
If you emailed the review package to the reviewers and they
emailed the package back to you, it will appear in the Inbox.

3 If the review package does not appear in the Inbox:

□ Select **Review** > **Import Review Package**.

□ Select the review package.

□ Click **Open**.

4 Double-click a reviewed topic or snippet in the list.
The reviewed version of the file opens in the Flare editor.

5 Click **Accept Change** or **Reject Change** for each tracked
change from the reviewer(s), or click **Accept All Changes** or
Reject All Changes to accept/reject all changes in a file.

6 Click 🖫 to save the reviewed file.

7 To replace the pre-reviewed file with the approved reviewed
version, highlight the file in the File Reviews window and click
🗎.

Sending topics for review in Central 🆕

You can also send topics for review in Central. With this approach,
reviewers do not need to install Contributor. Instead, they use
Central's web-based review features to add tracked changes and/or
annotations to topics.

In Flare 2023 and later, you can create a review package to send one
topic or multiple topics to reviewers. If your topics contain snippets,
the snippets will be automatically included in the review package.

If you send your topics using the TOC, Flare will include the TOC and all
of the topics that are linked to by the selected book and its sub-books
and pages.

◇ *Reviewers will need a Central user account to review topics in
Central.*

Shortcut	Tool Strip	Ribbon
Alt+R, S	⬚✉ (Review toolbar)	Review > Send for Review

To send topics for review in Central:

1 Select **Review** > **Send for Review**.

  The Send Files for Review wizard appears. Click ⬚.
  The Open File dialog box appears.

2 Select the topics you want to send for review and click **Open**.

  ◇ *If a selected topic contains a snippet, the snippet will also be added to the review package.*

3 Select a **TOC** to include with your review package.

4 Click **Next**.

5 Select **Send to MadCap Central**.

6 If you have defined multiple definitions for your variables and selected different definitions for your targets, select a target to specify the variable definitions that will appear in your review topics.

7 If you use condition tags in the selected topics, click ⬚ to include or exclude content in your review topics based on their condition tags.

8 Click **Next**.

9 Select the **Reviewers**.

10 Click **Send**.

To send topics for review in Central using the TOC:

1 Open a TOC.

2 Right-click a book and select **Send for Review**.
  The Send Files for Review wizard appears.

3 Select **Send to MadCap Central**.

4 Type a **Review Package Name**.

5 Type a **Review Package Description**.

6 Click **Next**.

7 Select the reviewer(s).

8 Click **Send**.
The files are uploaded to Central for review and listed in your Sent tab, and the reviewers are sent a review notification email.

Accepting review edits from Central

After you close a topic in review, you can accept or reject the review comments and accept the reviewed topic. When you accept the reviewed topic, the original topic is replaced by the reviewed topic.

Shortcut	Tool Strip	Ribbon
Alt+R, T	(Review toolbar)	Review > Topic Reviews

To accept reviewed topics from Central:

1 Select **Review** > **File Reviews**.
The File Reviews window appears.

2 In the text box at the top of the window, select **Inbox**.

3 Select the file(s) you want to close for review.

   *You can hold **Shift** or **Ctrl** to select multiple files.*

4 Click .

5 Double-click a closed file in the list.
The reviewed version of the file opens in the XML Editor.

6 Click **Accept Change** or **Reject Change** for each tracked change from the reviewer(s), or click **Accept All Changes** or **Reject All Changes** to accept/reject all changes in a file.

7 Click to save the reviewed file.

8 To replace the pre-reviewed file with the approved reviewed version, highlight the file in the File Reviews window and click ⧉.

Reviewing a list of topics out for review

You can review a list of topics that have been sent for review.

To view a list of topics that have been sent for review:

1 Select **Review** > **File Reviews**.

2 In the text box at the top of the window, select **Sent Files**.

Setting review display options

You can specify how track changes and annotations appear in the XML Editor.

To set the review display options:

1 Select **File** > **Options**.
The Options dialog box appears.

2 Select the **Review** tab.

3 Type your **User Name**.

4 Type your **Initials**.

5 Select a **Tracked Changes Display** option.

6 If you want to include change bars, set the **Change Bar** option to **On**.

7 Select a **User Colors** option.

Pulse

You can use MadCap Pulse to add user commenting to HTML5 and WebHelp targets. In addition to adding comments, users can like topics, ask and answer questions, and post files, articles, and images.

Setting up Pulse

You can install Pulse on the same server as your HTML5 or WebHelp content (the recommended approach), or you can install Pulse on a separate server. The Pulse server will need:

- Microsoft Windows Server 2008 or later
- Microsoft .NET Framework 4.0 or 4.5 (recommended)
- Microsoft SQL Server 2008 Standard or later
- Microsoft Internet Information Server (IIS) 7 or later
- Microsoft ASP.NET 4.0 or later

Pulse user levels

Pulse provides four user levels: unregistered ("general public"), basic ("customer"), advanced ("employee"), and administrator. The following table summarizes the tasks that each type of user can perform:

Task	Unregistered	Basic	Advanced	Admin
Read comments	✓	✓	✓	✓
Search comments	✓	✓	✓	✓
See ratings and interactions	✓	✓	✓	✓
Rate topics		✓	✓	✓
Create user profiles		✓	✓	✓
Post status updates		✓	✓	✓
Comment on, like, and follow posts		✓	✓	✓

Task	Unregistered	Basic	Advanced	Admin
Upload files		✓	✓	✓
Subscribe to topics		✓	✓	✓
Delete and edit their posts		✓	✓	✓
Assign tasks			✓	✓
Message users			✓	✓
Mark favorites			✓	✓
Vote on answers			✓	✓
Filter and view files			✓	✓
Access groups			✓	✓
Delete posts or files				✓
Modify server settings				✓
View reports				✓

Enabling Pulse in a skin

You can set up a skin to include the Pulse community tab and display user comments at the bottom of your topics.

To enable Pulse in a skin:

1 Open a skin.
The Skin Editor appears.

2 Select the **Community** tab.

3 Select the **Display topic comments at the end of each topic** option.

4 Select the **Display Community Search Results** option.
The total number of search results will display in parenthesis beside the Community Results heading in the search results.

Enabling Pulse in a target

You can add Pulse to HTML5, WebHelp, or WebHelp Plus targets.

To enable Pulse for a target:

1 Open a target.

2 Select the **Analytics** tab.

3 For **Provider**, select **Pulse/Feedback Server**.

4 Type your Pulse server's **URL**.

5 Save the target.

Creating a Pulse profile

Pulse users can create a profile themselves to interact with topics and other users, or they can be invited to join the Pulse community by the administrator.

To create a Pulse profile:

1 Open an HTML5 or WebHelp target that uses Pulse.

2 Select the **Community** tab or open a topic.

3 Click **Register**.

4 Type your **First Name** and **Last Name**.

5 Type your **Email Address**.

6 Type and confirm your **Password**.

7 Click **Register**.
You will receive a registration email.

8 Open your email and complete the registration.

Editing your Pulse profile

You can edit your Pulse avatar, contact information, and notification options, such as whether you receive an email when other users comment on your posts.

To edit your profile:

1 Open an HTML5 or WebHelp target that uses Pulse.

2 Select the **Analytics** tab.

3 Click **Edit My Profile**.
The Settings window appears.

4 Select the **Personal** tab to change your information or select the **Notifications** tab to change your notification options.

Viewing notifications

You can use the Notifications window to view summary of your posts, tasks, questions, and groups, or any activities for people you are following.

To view notifications:

1 Open an HTML5 or WebHelp target that uses Pulse.

2 Select the **Analytics** tab.

3 Click **Notifications**.
The Notifications window appears.

Viewing a report

MadCap Pulse provides the following reports:

- Browser statistics
- Context-sensitive help calls
- Most active groups
- Most commented activities
- Most commented file shares
- Most commented image shares
- Most liked link shares
- Most liked people

- Operating system statistics
- Overall activity
- Search phrases
- Search phrases with no results
- Storage usage
- User activity
- User count
- Viewed topics

To view a report:

1 Open your Pulse Admin site in a browser.

2 Click **Administration**.

3 Select **Reports**.
A list of reports appears.

4 Click a report.

5 If you have multiple Pulse communities, you can filter the reports by community.

Source control

You can add (or "bind") your project to any source control application that supports the Microsoft Source Code Control API (SCC API). A source control application can prevent team members from overwriting each other's changes, save past versions of files, and identify changes made to files (and who made them).

Binding a project to source control

Flare provides built-in support for Git, Perforce Helix Core, Subversion (SVN), and Team Foundation Server (TFS). With these applications, you can bind your project and check in/check out files within Flare (or push/pull files if you are using Git). For other source control applications, you will need a plug-in to integrate the application with Flare. Or, you can check out the files in your source control application, use Flare to make changes, then check in the files outside of Flare.

Shortcut	Tool Strip	Ribbon
Alt+P, R	(Project toolbar)	Project > Project Properties

To bind a project to a source control application:

1   Select **Project** > **Project Properties**.
    The Project Properties dialog box appears.

2 Select the **Source Control** tab.

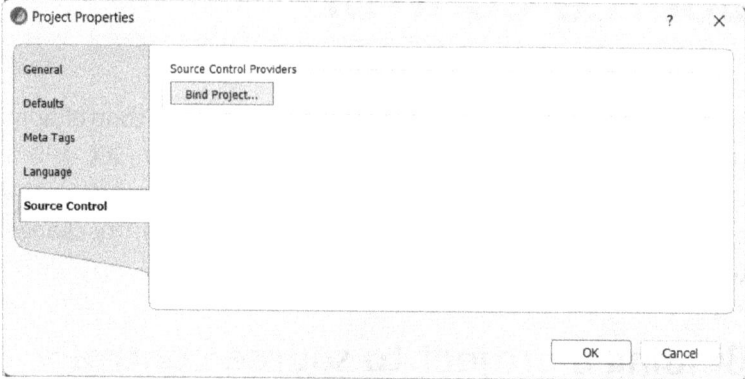

3 Click **Bind Project**.

The Bind Project dialog box appears.

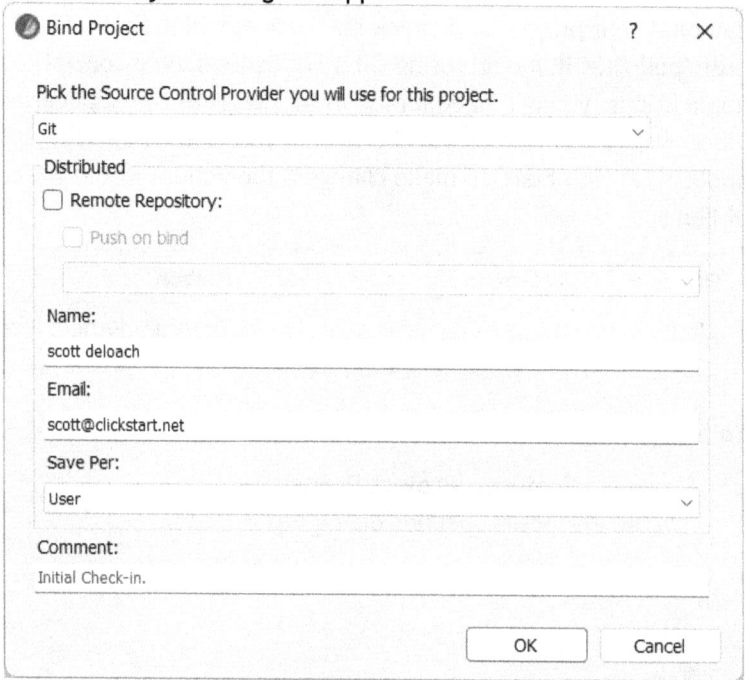

4 Select a **Source Control Provider** and provide the required information.

5 Click **OK**.

Importing a project from source control

If you work on a team, one team member should bind the project to add it to source control. The other team members need to import the project from source control. In other words, binding the project copies it from one team member's PC to the source control server. The other team members do the opposite: they copy the files from the source control server to their PC.

There are two ways to import a project directly from source control. You can import the project, or you can copy the project from a server and set Flare to detect that it is bound to source control.

To import a project from source control:

1 Select **File** > **New Project** > **Import Project**.
The Import Project from Source Control wizard appears.

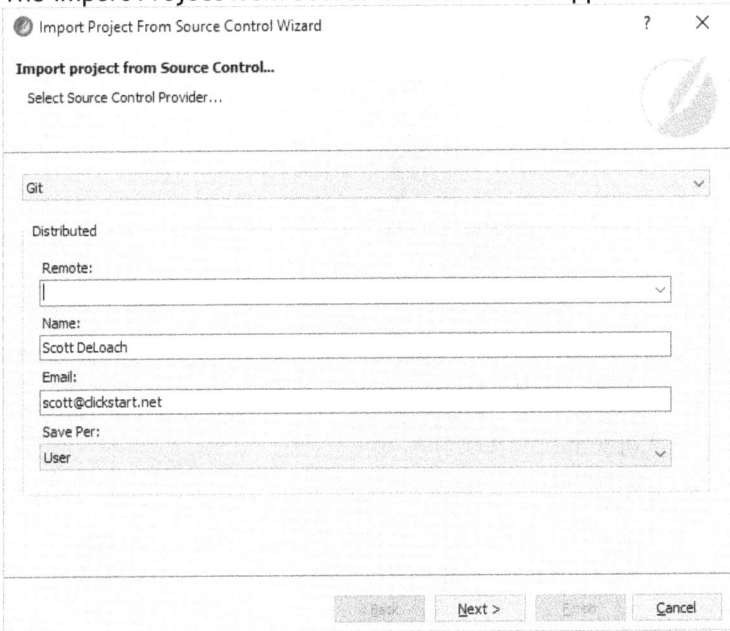

2 Select a **Source Control Provider**.

3 Select the same server, team project, and other options that were used to bind the project.

4 Click **Finish**.

To use bind detection:

1 Select **File** > **Options**.

The Options dialog box appears.

2 Select the **Source Control** tab.

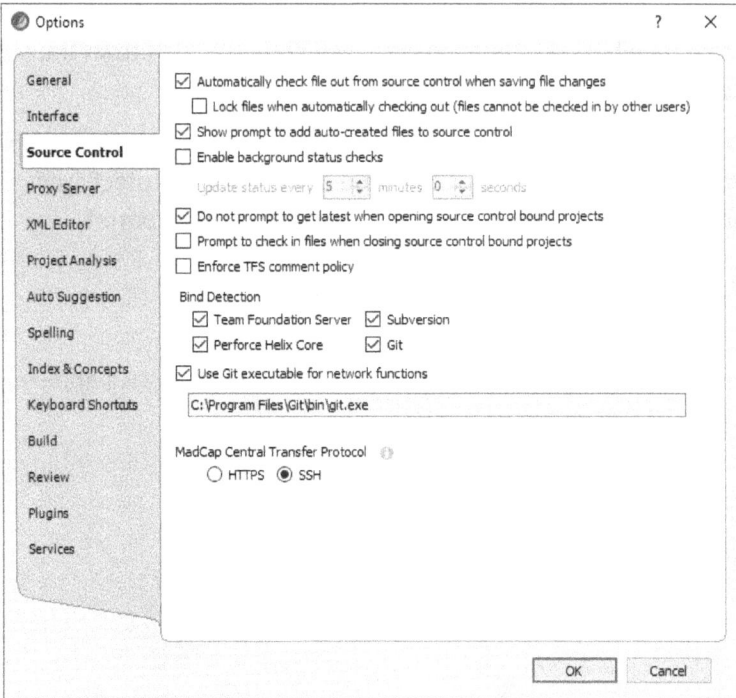

3 In the **Bind Detection** section, select the checkbox for your source control provider.

4 Copy the project files to your PC.

5 Open the project.

Flare will scan the network to find the source control binding. This step could take a considerable amount of time.

Adding a Git ignore file

By default, Git will bind all of the files in a project. If you bind a project in Flare, Flare will automatically create and setup an ignore file to prevent content in the following folders from being bound:

- ☐ Analyzer

- ☐ FileSync

- ☐ Output

- ☐ Project/Users

If you do not bind your project to Git using Flare, you should create an ignore file. MadCap recommends excluding the folders listed above.

To add an ignore file:

1 Select **View** > **Source Control Explorer**.

2 Click **Settings**.

3 Click **Add Ignore File**.
The new ignore file opens in the text editor.

Editing a Git ignore file

You can edit the ignore file to change which files are bound to source control.

To edit an ignore fle:

1 Select **View** > **Source Control Explorer**.

2 Click **Settings**.

3 Click **Edit Ignore File**.
The ignore file opens in the text editor.

4 Edit the ignore file.

5 Click **Save**.

Pulling files

Before you make changes to a file, you should make sure you have the latest version of the file from source control. In Git, this is called a "pull."

Shortcut	Tool Strip	Ribbon
Alt+S, G, L	(Standard toolbar)	Source Control > Pull

To pull the latest files from source control:

☐ Select **Source Control > Pull**.

Adding a file to source control

When you add a file to your project, you should also add ('push") it to the source control server. A ✛ plus sign icon will appear beside new files that have not been added to source control.

Shortcut	Tool Strip	Ribbon
Alt+S, A	(Standard toolbar)	Source Control > Add

To add a file to source control:

1 Select a file or folder that is not in source control.

2 Select **Source Control > Add**.
—OR—
Right-click the file or folder and select **Source Control > Add**.
The Commit dialog box appears.

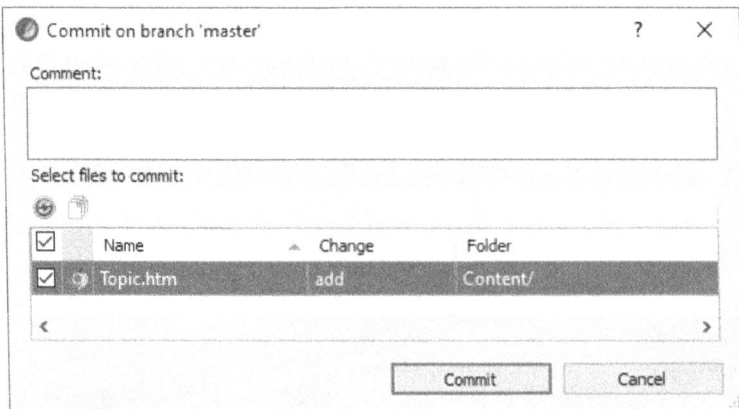

3 Type a **Comment**.

4 Click **Commit**.

Committing changes

After you have modified the local version of a file, you can commit your changes. When you commit, you can add a comment that describes the changes. Each commit it archived in source control, and you can undo (rollback) the changes later if needed.

A ✓ checkmark icon appears beside modified files that have not been pushed to source control.

Shortcut	Tool Strip	Ribbon
Alt+S, C, O	⬛ (Standard toolbar)	Source Control > Commit

To commit changed files:

1 Select **Source Control > Commit**.

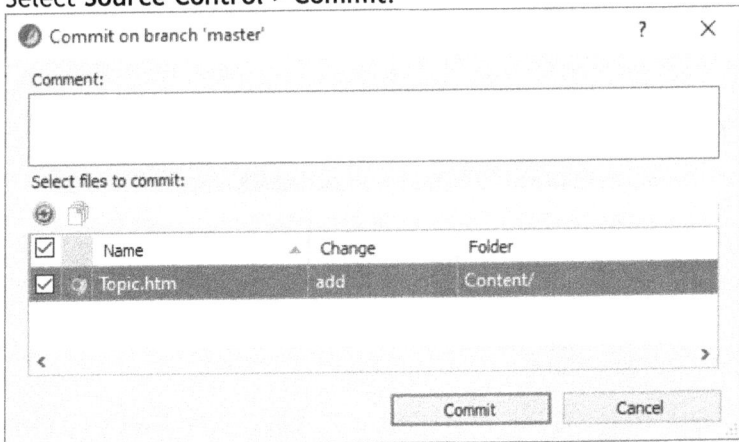

2 Type a **Comment**.

3 Click **Commit**.

Pushing files

After you have committed your changes, you can push the changed files to source control so that other users can work with them.

Shortcut	Tool Strip	Ribbon
Alt+S, C, I	(Standard toolbar)	Source Control > Check In

To push committed changes:

1 Select **Source Control > Push**.
The Select Remote for Push dialog box appears.

2 Select a remote repository.

3 Click **OK**.

Viewing a list of pending changes

You can view a list of files that you have modified or added but have
not pushed to the source control server.

Shortcut	Tool Strip	Ribbon
Alt+S, P+C	(Standard toolbar)	Source Control > Pending Changes

To view a list of pending changes:

1 Select **Source Control > Pending Changes.**
The Pending Changes pane appears.

2 Scroll to the right to view the Status and User columns.

3 To sort the list, click a column heading.

Viewing differences between versions

You can compare any two versions of a file to review how a file has
changed.

Shortcut	Tool Strip	Ribbon
Alt+S, S	(Standard toolbar)	Source Control > Show Differences

To view differences between versions:

1 Select a file.

2 Select **Source Control** > **Show Differences**.
The History dialog box appears.

3 Select two versions of the file.

4 Click **Show Differences**.

Rolling back to a previous version

You can roll back to a previous checked in version of a file.

Shortcut	Tool Strip	Ribbon
Alt+S, V	▣ (Standard toolbar)	Source Control > View History

To roll back to a previous version:

1 Select a file or folder.

2 Select **Source Control** > **View History**.
—OR—
Right-click and select **Source Control** > **View History**.
The History dialog box appears.

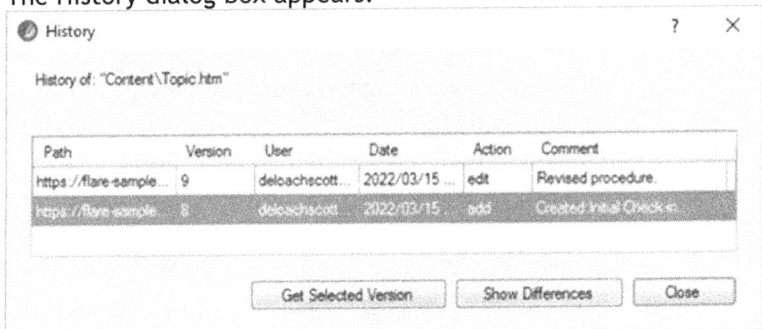

3 Select a version of the file.

4 Click **Get Selected Version**.
The selected version will be copied to the project.

5 Click **Close**.

Merging changes

If two users have modified a file at the same time, you can merge the changes and try to preserve both users' modifications.

Shortcut	Tool Strip	Ribbon
Alt+S, C, I	⌨ (Standard toolbar)	Source Control > Check In

To merge changes:

1 Check in or get the latest version of a file.
   If your version is different from the version in source control, the Resolve Conflicts dialog box appears.

2 Click **Auto Merge All**.
   If the differences do not conflict, the versions are merged. If the differences conflict, complete the following steps.

3 Click **OK**.

4 Click **Resolve**.

5 Select a resolution option:

   □ **Merge changes in merge tool** — Opens a merging interface, which lets you see exactly what changes were made and choose which to keep.

   □ **Undo my local changes** — Automatically removes your changes and keeps changes from other authors.

   □ **Discard server changes** — Automatically removes changes from other authors and keeps your changes.

6 If you selected **Merge changes in merge tool**, you can:

   □ **Right-click** to open a context menu that can be used to keep or ignore a change to the server or local version.

   □ **Left-click** to keep the change to the server or local version.

   □ **Type** content to edit the content and merge the versions yourself.

7 Click **OK** when all of the conflicts are resolved.

Creating a branch

If you are using Git for source control, including Git as part of MadCap Central, you can create a branch to isolate changes. For example, you could create a new branch for changes you are making for a new version of a product. Later, you can merge these changes into your main branch.

To create a branch:

1   Commit any pending changes on the current branch.

2   Select **Source Control** > **Branch**.
    —OR—
    Click the current branch's name in the bottom right corner Flare window.
    The Branch Management dialog box appears.

3   Click **Create**.
    The Create Branch dialog box appears.

4   Select a **Source Branch**.

5   Type a **Branch Name**. Branch names cannot include spaces.

6   If you want to switch to the new branch, select **Switch to branch**.

7   Click **Create**.

8   Click **Close**.

Getting a branch

You can get a remote branch that a coworker has created. Once you get a remote branch, you can switch to it and contribute your own changes.

To get a branch:

1   Select **Source Control** > **Pull** to update the list of remote branches.

2 Select **Source Control > Branch**.
—OR—
Click the current branch's name in the bottom right corner Flare window.
The Branch Management dialog box appears.

3 Select the **Remotes** tab.

4 Select a branch.

5 Click **Switch**.

6 Click **Close**.

Switching to a branch

You can switch to a branch to make changes to the files in the branch.

To switch to a branch:

1 Commit any pending changes on the current branch.

2 Select **Source Control > Branch**.
—OR—
Click the current branch's name in the bottom right corner Flare window.
The Branch Management dialog box appears.

3 Select the **Locals** or **Remotes** tab.
Switching to a remote branch will also add it to the Locals tab.

4 Select a branch.

5 Click **Switch**.

6 Click **Close**.

Merging a branch

You can merge the changes made in a branch with another branch, or more likely, with the "main" branch.

To merge a branch:

1   If needed, switch to the branch into which you want to merge.

2   Select **Source Control** > **Merge**.
    The Select Branch to Merge dialog box appears.

3   Select the branch you want to merge into the current
    ("active") branch.

4   Click **OK**.
    The selected branch is merged into the current branch.

5   Resolve any merge conflicts (if they exist).

Reverting a branch

You can revert (or "back out") a commit within a branch. Reverting
allows you to easily undo past changes.

To revert a branch:

1   If needed, switch to the branch you want to revert.

2   Select **Source Control** > **Branch History**.
    The Branch History dialog box appears.

3   Select the commit you want to revert.

4   Click **Revert**.

5   Click **Accept**.

6   Click **Close**.

Deleting a branch

You can delete a branch if you no longer need the changes that were
made in the branch. For example, you might delete a branch you
created for testing.

To delete a branch:

1 If you want to delete the current (active) branch, switch to another branch.

2 Select **Source Control > Branch**.
—OR—
Click the current branch's name in the bottom right corner Flare window.
The Branch Management dialog box appears.

3 Select a branch.

4 Click **Delete**.
The Delete Branch dialog box appears.

5 Specify whether you also want to delete the matching local or remote branch (if it exists) and click **OK**.

6 Click **Close**.

Disabling a source control provider for a project

Flare has built-in support for Git, Perforce Helix Core, Subversion, and Team Foundation Server. When you bind a project to one of these source control providers, the source control commands appear in the Flare UI when you open the project. If you plan to use a separate source control client, you can disable your source control provider and hide the source control commands in Flare.

To disable a source control provider for a project:

1 Select **Project > Project Properties**.

2 Select the **Source Control** tab.

3 Clear the **Enabled** checkbox.

4 Click **OK**.

5 Close and reopen the project. The source control commands will disappear.

Disabling a source control provider for all projects

If you use another source control client, you can disable the source control provider in Flare and hide the source control commands in the Flare UI.

To disable a source control provider for all projects:

1 Select **File** > **Options**.

2 Select the **Source Control** tab.

3 Clear the checkbox for the source control provider you want to disable.

4 Click **OK**.

5 Close and reopen the project. The source control commands will disappear.

Disconnecting from source control

If you need to work on a project while you are not connected to the source control server, you can disconnect from source control and reconnect when you have access to the server again. This option might be useful if you need to work at home.

✎ *Disconnecting is only available with Subversion, Team Foundation Server, and Perforce Helix Core. It is not supported by Git.*

Shortcut	Tool Strip	Ribbon
Alt+S, D	⬚ (Standard toolbar)	Source Control > Disconnect

To disconnect from source control:

1 Check out the files you need to modify.

2 Select **Source Control** > **Disconnect**.
   The commands in the Source Control ribbon are disabled.

To reconnect to source control:

1   Select **Source Control > Reconnect**.
    The commands in the Source Control ribbon are enabled.

2   Check in the files you changed while you were disconnected.

Unbinding from source control

If you need to move your project to another source control server, you can unbind it from source control.

Shortcut	Tool Strip	Ribbon
Alt+P, R	(Project toolbar)	Project > Project Properties

To unbind a project from source control:

1   Select **Project > Project Properties**.
    The Project Properties dialog box appears.

2   Select the **Source Control** tab.

3   Click **Unbind Project**.

Source control with Central

MadCap Central is a cloud-based product that can be used to manage, build, and host the output from your Flare projects. Central is fully integrated into Flare, and it can be used as a source control system or with another source control system. Central's source control is based on Git. For more information about source control, see "Source control" on page 425.

Logging in to Central NEW!

If you have a Central account, you can log in to Central from Flare. Once you log in, you can upload ("bind") projects, download ("import") projects, and use Central's source control features, such as pushing and pulling changed files.

◇ *All versions of Flare support MadCap Central servers in the United States. Flare 2023 and later versions also support Central servers in Europe.*

Shortcut	Tool Strip	Ribbon
Alt+V, M, C	View > MadCap Central	View > MadCap Central

To log in to Central:

1 Select **View** > **MadCap Central**.
   The MadCap Central pane appears.

2 Select a **Server Location**.

3 Type your **Username**.

4 Type your **Password**.

5 Click **Login**.

Binding a project to Central

If you bind a project to Central, you can use Central as a source control system. Once your project is in Central, any Central user who has permission to access the project in Central can also build and publish targets without needing Flare.

◇ *To bind a project, you will need to have permission to "Upload New Projects" in Central.*

Shortcut	Tool Strip	Ribbon
Alt+V, M, C	View > MadCap Central	View > MadCap Central

To bind a project to Central:

1 Select **View > MadCap Central.**

2 If you are not logged in to Central, log in.

3 Click .
The Bind Project dialog box appears.

4 If needed, select a **License** and/or change the **Project Name** and/or **Project Description**.

Importing a project from Central

If a coworker binds a project to Central, you can import it to work with the project.

◇ *To import a project from Central, you will need to have permission to "Import/Pull" projects in Central.*

Shortcut	Tool Strip	Ribbon
Alt+V, M, C	View > MadCap Central	View > MadCap Central

To import a project from Central:

1 Select **View > MadCap Central.**

2 If you are not logged in to Central, log in.

3 Click .

The Import Project dialog box appears.

4 For **My Projects**, select the project you want to import.

5 For **Destination**, click browse and select the folder where you want to store your project.

6 Click **OK**.

7 When the import is finished, click **Open** to open the project in Flare.

Pushing changes to Central

If you modify or add content, you should push your changes to Central. Once you push your changes to Central, your coworkers can pull them to their local repository.

◇ *To push changes to Central, you will need to have permission to "Push" files in Central.*

Shortcut	Tool Strip	Ribbon
Alt+V, M, C	View > MadCap Central	View > MadCap Central

To push changes to Central:

1 Select **View** > **MadCap Central**.

2 If you are not logged in to Central, log in.

3 Select **Source Control** > **Push**.

If you did not commit the changes before starting the push, the Commit dialog box appears. Click **Yes** and click **Commit**.

The Select Remote for Push dialog box appears.

4 Select a **Remote** repository.

5 Click **OK**.

Pulling changes from Central

If you work as a team, you should pull your coworkers changes to your repository to update your copy of the project.

◇ *To pull changes from Central, you will need to have permission to "Import/Pull" files in Central.*

Shortcut	Tool Strip	Ribbon
Alt+V, M, C	View > MadCap Central	View > MadCap Central

To pull changes from Central:

1 Select **View** > **MadCap Central**.

2 If you are not logged in to Central, log in.

3 Select **Source Control** > **Pull**.
The Select Remote for Pull dialog box appears.

4 Select a **Remote** repository.

5 Click **OK**.

Logging out of Central

Usually, you won't need to log out of Central. However, you can log out if you need to log in again as a different user or for a different company or team.

Shortcut	Tool Strip	Ribbon
Alt+V, M, C	View > MadCap Central	View > MadCap Central

To log out of Central:

1 Select **View** > **MadCap Central**.

2 Click .

Unbinding projects from Central

If you unbind a project, you can edit the project locally, but you cannot push any changes to Central.

◇ *To unbind a project, you will need to have permission to "Upload New Projects" in Central.*

Shortcut	Tool Strip	Ribbon
Alt+V, M, C	View > MadCap Central	View > MadCap Central

To unbind a project from Central:

1 Select **View** > **Central**.

2 If you are not logged in to Central, log in.

3 Click .
The Confirmation window appears.

4 Click **OK**.

SharePoint integration

You can import SharePoint files into your Flare project. For example, you can reuse PDFs, logos, or other documents that are stored in a company-wide SharePoint directory.

TIP *You can also add files from SharePoint to your project as external resources and publish targets to SharePoint by making SharePoint as publishing destination. For more information, see "Importing external resources" on page 105 and "Publishing a target" on page 348.*

Connecting to a SharePoint server

If you connect your project to a SharePoint server, you can include files from SharePoint in your project.

Shortcut	Tool Strip and Ribbon
Alt+P, H	View > SharePoint Explorer

To connect to a SharePoint server:

1 Select **View > SharePoint Explorer**.
   The SharePoint Explorer pane appears.

2 Click 🖫.
   The Connect to SharePoint Server dialog box appears.

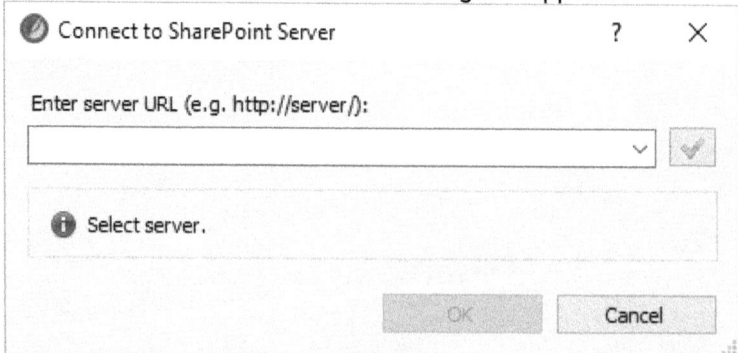

3 Type the path to the SharePoint server.

4 Click ![checkmark]to validate the server's URL.

5 Click **OK**.
The SharePoint Explorer pane appears.

Copying and mapping SharePoint files

After you connect to a SharePoint server, you can add files from SharePoint to your project. If you map the files, you can synchronize any changes that are made to your local copy of the file or the version in SharePoint.

To copy and map SharePoint files:

1 Select **View** > **SharePoint Explorer**.
The SharePoint Explorer pane appears.

2 Right-click a file and select **Copy to Project**.

3 Click **OK**.
The Copy to Project dialog box appears.

4 Select **Keep file synchronized (create mapping)**.

5 Click **OK**.
An orange icon appears beside the file in the Content Explorer.

Synchronizing SharePoint files

When you synchronize a file, Flare compares your local version of the file to the version in SharePoint. If they are different, you can update either version.

Shortcut	Tool Strip	Ribbon
Alt+P, Y	Tools > Synchronize Mapped Files	Project > Synchronize Files

To synchronize SharePoint files:

1   Select **Project** > **Synchronize Files**.
    The Synchronize Files dialog box appears.

2   Select one of the following options:

Option	Description
Synchronize Files	Compares the version of the files in your project to the version in SharePoint and updates the version that is out of date.
Import Files	Copies out-of-sync SharePoint files into your project, even if the files in your project have been modified more recently that the versions in SharePoint.
Export Files	Copies out-of-sync files in your project to SharePoint, even if the files in SharePoint have been modified more recently that the versions in your project.
Custom	Allows you to manually select which files you want to import or export.

3   Click **Synchronize**.

4   Click **OK**.

Checking out a file from SharePoint

If you check out a file, other users are not allowed to make changes to the file until you check it in again.

Shortcut	Tool Strip and Ribbon
Alt+F, H, O	File > SharePoint > Check Out

To check out a file:

1   Select the file(s) you want to check out.

2   Select **File** > **SharePoint** > **Check Out**.
    —OR—
    Right-click the file(s) and select **Check Out**.

3   Click **Check Out**.

Checking in a file to SharePoint

When you check in a file to SharePoint, other users are allowed to make changes to the file.

Shortcut	Tool Strip and Ribbon
Alt+F, H, I	File > SharePoint > Check In

To check in a file:

1   Select the file(s) you want to check in.

2   Select **File** > **SharePoint** > **Check In**.
    —OR—
    Right-click the file(s) and select **Check In**.

3   Click **Check In**.

Archive project

You can zip or export your project to archive it. Zipping a project is faster, but exporting a project allows you to specify which files are included in the archive.

Zipping a project

You can zip your project to archive it to a server or to send it to someone. When you zip a project in Flare, Flare creates an flprjzip file. Unlike a standard zip file, an flprjzip file can be emailed without the included JavaScript files being removed by an email filter.

Shortcut	Tool Strip and Ribbon
Alt+P, Z	Project > Zip Project

To zip a project:

1 Select **Project > Zip Project**.
   The Create Package dialog box appears.

2 Type or select a package file path and name.

3 Click **Create**.
   Flare creates the flprjzip file.

Exporting a project

You can export your project to archive all or selected files in a large project or to send selected files for translation.

Shortcut	Tool Strip and Ribbon
Alt+P, E, P	Project > Export Project

To export a project:

1 Select **Project** > **Export Project**.
The Export Project wizard appears.

2 Type a **New Project Name** for the exported project.

3 Type or select a **New Project Path** for the exported project.

4 For **Export From**, select whether you would like to export the entire project or only part of the project.

5 For **Output**, select whether you would like to create project files, a zip file, or a project template.

6 Click **Next**.

7 Select whether you want to convert variables to text.

8 Select whether you want to convert snippets to text.

9 Click **Finish**.

Sample questions for this section

1 Topic templates have the following extension:
A) htm
B) htt
C) temp
D) fltt

2 File tags can be applied to:
A) Topics only
B) Topics and images
C) Topics, images, videos, and sounds
D) Any type of file in your project

3 Meta tags can be used to:
A) Specify author or status levels for files in your project
B) Include "hidden" information inside the head section of HTML5 topics
C) Group topics together to create "meta" links
D) Organize topics

4 You can create reports based on the following information in Flare:
A) File tags
B) Used and unused styles, variables, and images
C) Both of the above
D) None of the above

5 Annotations and tracked changes are included when you send a topic for review in Contributor.
A) True
B) False

6 Reviewers need to install Flare to review your topics.
A) True
B) False

7 Flare provides built-in support for which source control applications?
A) Flare does not provide build-in support for source control. You must use an external client.

B) Subversion (SVN) and CVS

C) Git, Perforce Helix Core, SVN, and TFS

D) All of them

8 What is MadCap Central?

A) An online discussion group where users can discuss Flare and other MadCap products.

B) A cloud-based platform for content management, hosting, and task tracking.

C) A suite of products, including Flare, Mimic, and Capture.

D) A user conference about MadCap products.

eLearning

This section covers:

- ☐ Multiple choice questions
- ☐ Multiple response questions
- ☐ Test nodes
- ☐ eLearning toolbar proxies
- ☐ eLearning toolbar skin components
- ☐ Test results proxies
- ☐ Test results skin components
- ☐ xAPI and SCORM packages
- ☐ PDF test keys

Creating a multiple choice question

You can add a multiple choice question (or even multiple questions) to a snippet or topic.

Shortcut	Tool Strip	Ribbon
Alt+L, M, C	none	eLearning> Multiple Choice

To create a multiple choice question:

1 Open a topic.

2 In the XML Editor, position your cursor where you want to insert the multiple choice question.

3 Select **eLearning** > **Multiple Choice**.
   A sample question and answer is added to the topic.

4 Type the question.

5 Type the first answer.

6 To add more answers, press **Enter** and type the answer.

7 Click the circle before the correct answer.

Creating a multiple response question

You can add a multiple choice question (or even multiple questions) to a snippet or topic.

Shortcut	Tool Strip	Ribbon
Alt+L, M, C	none	eLearning> Multiple Response

To create a multiple response question:

1 Open a topic.

2 In the XML Editor, position your cursor where you want to insert the multiple response question.

3 Select **eLearning** > **Multiple Response**.
   A sample question and answer is added to the topic.

4 Type the question.

5 Type the first answer.

6 To add more answers, press **Enter** and type the answer.

7 Click the square before each correct answer.

Creating a question using existing content

You can create a question based on existing paragraphs of content. The first paragraph will become the question and the following paragraphs will become the answers.

To create a question using existing content:

1 Open a topic.

2 Select the paragraphs that contain the question and answers.

3 Select **eLearning** > **Multiple Choice** or **Multiple Response**.

4 Click a circle (multiple choice) or square(s) to select the correct answer(s).

Switching the question type

You can change a multiple choice question to a multiple response question or a multiple response question to a multiple choice question.

To switch the question type:

1 Open a topic that contains a question.

2 Right-click inside a question and select Switch to Multiple Response or Switch to Multiple Choice.

3 Click a circle (multiple choice) or square(s) to adjust the correct answer(s) if needed.

Adding feedback to a question

You can add correct and incorrect answer feedback for a question. The feedback can contain any type of content, including text, images, tables, and links. The feedback can appear after the question has been answered and on the test results page once the test has been completed.

To add feedback to a question:

1   Open a topic that contains a question.

2   Click inside the question.

3   Select **eLearning** > **Add Feedback**.
    Sample correct and incorrect answers are added below the question.

4   Add the correct answer feedback.

5   Add the incorrect answer feedback.

6   If you want to allow the user to see the feedback after they answer the question (rather than at the end of the quiz), click **Add Submit Button.**

Adding a test node to a TOC

If you are creating a test, you should add a test node to your TOC. The test node should include your test topics as sub items (like a TOC book). If you are creating ungraded "knowledge checks," you do not need to add a test node.

To add a test node to a TOC:

1   Open a TOC.

2   Select **eLearning** > **Create Test**.
    A test node is added to the TOC.

3   If needed, move the test node to a new location in the TOC.
    Test nodes are often the last item in the TOC.

4 Right-click the test node and select **Test Options**.
The Properties dialog box appears.

5 Specify whether you want to Randomize the test answers.

6 Specify whether you want to **Limit Test Retries**.

7 Specify the **Passing Score**.

8 If you want to use a custom pass and/or fail page, click .
The custom pass/fail page should contain a test results proxy.

9 Click **OK**.

10 Add the test topics as pages to the test node in the TOC.

Adding an eLearning toolbar proxy

You should add an eLearning toolbar to allow the user to move to the next or previous question. It can also include a progress indicator.

To add an eLearning toolbar proxy:

1 Open the template page or topic that will contain the eLearning toolbar proxy.

2 Position your cursor where you want to insert the proxy.

3 Select **Insert** > **Proxy** > **eLearning Toolbar Proxy**.
The eLearning Toolbar Proxy dialog box appears.

4 Click **OK**.

Creating an eLearning toolbar skin component

You can create an eLearning toolbar skin component to add and format a previous button, next button, and progress indicator.

To create an eLearning toolbar skin:

1 Select **File** > **New**.
—OR—

Right-click the **Skins** folder and select **Add Skin**.
The Add File dialog box appears.

2   For **File Type**, select **Skin**.

3   For **Source**, select **HTML5 Component - eLearning Toolbar**.

4   Type a **File Name**.
Skins have a .flskn extension. If you don't type the extension, Flare will add it for you.

5   Click **Add**.
The skin appears in the Skins folder in the Project Organizer and opens in the Skin Editor.

Designing an eLearning Toolbar skin component

You can specify the fonts, colors, and spacing of the eLearning toolbar buttons and progress indicator.

To design an eLearning Toolbar skin component:

1   Open a test results skin component.

2   Select the **Setup** tab.

3   Select the buttons you want to include in the toolbar.

4   Select the **Styles** tab.

5   Modify the styles as needed. You can change color, font, border, margins, padding, and background settings for the toolbar and progress bar.

6   Select the **UI Text** tab.

7   Modify any of the labels or tooltips.

Applying an eLearning Toolbar skin component to a proxy

You can apply an eLearning skin component to an eLearning toolbar proxy, or you can select an eLearning skin component in a target. If you apply the skin component to the proxy, it will automatically be used for every target that includes the proxy.

To apply an eLearning toolbar skin component to eLearning toolbar proxy:

1 Open a template page or topic that contains the eLearning Toolbar proxy.

2 Right-click the eLearning toolbar proxy and select **Edit eLearning Toolbar** proxy.
   The eLearning Toolbar Proxy dialog box appears.

3 Select the **Skin File**.

4 Click **OK**.

Adding a test results proxy

You can create a custom correct or incorrect test results page to customize the test results page's appearance and content. A custom correct or incorrect page should include a test results proxy.

To add a test results proxy to a topic:

1 Open the topic that will contain the test results proxy.

2 Position your cursor where you want to insert the proxy.

3 Select **Insert** > **Proxy** > **Test Results Proxy**.
   The Test Results Proxy dialog box appears.

4 Click **OK**.

Creating a test results skin component

You can create a test results skin component to specify whether the following information is included on the test results page:

- ☐ Pass/Fail

- ☐ Score

- ☐ Questions

If the questions are included, you can also include or exclude the user's answers, the correct answers, and each question's feedback.

To create a test results skin component:

1 Select **File** > **New**.
—OR—
Right-click the **Skins** folder and select **Add Skin**.
The Add File dialog box appears.

2 For **File Type**, select **Skin**.

3 For **Source**, select **HTML5 Component - Test Results**.

4 Type a **File Name**.
Skins have a .flskn extension. If you don't type the extension, Flare will add it for you.

5 Click **Add**.
The skin appears in the Skins folder in the Project Organizer and opens in the Skin Editor.

Designing a test results skin component

You can format the test feedback area, including the score, pass/fail, and question feedback areas.

To design a test results skin component:

1 Open a test results skin component.

2 Select the **Setup** tab.

3 Select the test options you want to include on the results page.

4 Select the **Styles** tab.

5 Modify the styles as needed. You can change fonts, icons, labels, borders, paddings, and background settings and other style properties.

6 Select the **UI Text** tab.

7 Modify any of the labels or tooltips.

Applying a test results skin component to a proxy

You can apply a test results skin component to a test results proxy, or you can select a test results skin component in a target. If you apply the skin component to the proxy, it will automatically be used for every target that includes the proxy.

To apply a test results skin component to a Test Results proxy:

1 Open a topic that contains a test results proxy.

2 Right-click the test results proxy and select **Edit Test Results** proxy.
The Test Results Proxy dialog box appears.

3 Select the **Skin File**.

4 Click **OK**.

Generating an xAPI or SCORM package

You can generate an xAPI (also called "Tin Can") or SCORM package when you build an HTML5 target. After you build the HML5 target, you can upload the ZIP package to your LMS.

To generate an xAPI ('Tin Can") or SCORM package for an HTML5 target:

1 Open an HTML5 target.

2 Select the **eLearning** tab.

3   Select an eLearning **Standard**.

4   Type a course **Name**.

5   Type a course **Description**.

6   Type a unique course **ID**. For SCORM, the ID is the name of the manifest file. For xAPI, the ID is the course's URL.

7   If you are creating an xAPI package, modify the xAPI **ID** if needed.

8   Select a **Tracking** option: **Use Quiz Results** or **Use Course Completion**.

9   If the Tracking option is set to **Use Course Completion**, type a completion **Percentage**.

Creating a PDF test key

You can create a PDF target to use as a test key.

To create a PDF test key:

1   Open a PDF target.

2   Select the **General** tab.

3   Ensure the Primary TOC is set to the course's TOC.

4   Select the **Advanced** tab.

5   Select **Show correct answers for eLearning questions**.

6   Click **Build**.

Accessibility

This section covers:

- ☐ Keyboard access
- ☐ ARIA attributes
- ☐ Table headers, captions, and summaries
- ☐ Alt text for cross references and hyperlinks
- ☐ Alt text for images, videos, and equations
- ☐ Access for WebHelp without stylesheets
- ☐ PDF/UA tagging
- ☐ Accessibility warnings

Accessibility overview

Accessibility focuses on making content accessible to all users, including users who may have challenges moving, seeing, and/or hearing. Accessibility features are recommended by Section 508 of the U.S Government's Rehabilitation Act and the W3C's Web Content Accessibility Guidelines (WCAG). Not providing accessible content is often seen as a form a discrimination, and many countries are introducing legislation to require accessibility features.

The following table lists the basic accessibility guidelines.

Accessibility guideline	See page
Include page titles	71
Provide keyboard access	466
Provide ARIA attributes	467
Use headings and table header rows	467
Include table captions and summaries	467
Include alt text for cross references and hyperlinks	468
Include alt text for images	469
Include alt text for videos	470
Include alt text for equations	470
Automatically adding alt text	471
Support access when stylesheets are disabled	471
Include PDF/UA tags	469
Finding and fixing accessibility issues	472

Provide keyboard access

HTML5 targets include the following built-in keyboard access features:

- **Menus** — Press **Tab** to move between menu items and **Enter** to select a menu item.

- **Toolbar** — Press **Tab** to move between buttons in a toolbar and **Enter** or **Space** to select a toolbar button.

- **Search bar** — Press **Tab** to move between elements in a toolbar and **Enter** or **Space** to select a search bar element.

Provide ARIA attributes

The W3C's ARIA (Accessible Rich Internet Applications) specification defines textual descriptions to help screen readers interpret and present HTML elements.

Flare automatically adds scope and role attributes to table header rows. HTML5 targets automatically include ARIA attributes for interactive elements such as drop-down, expanding, and toggler links, breadcrumbs, and glossary popup links.

Adding a header row to a table

You can add a header row to indicate how the content is organized in a table. Header rows also allow you to use a table style to format the header separately from normal table "data" rows.

Shortcut	Tool Strip	Ribbon
Alt+B, O	Table > Table Properties	Table > Table Properties

To add a header row to a table:

1 Right-click inside a table and select **Table Properties**. The Table Properties dialog box appears.

2 Select the **General** tab.

3 For **Number of header rows**, select **1**.

4 Click **OK**.

Adding a caption and summary to a table

You can add table captions and summaries to increase accessibility. A table caption provides a literal description of the table, such as

"Monthly average temperatures in Bermuda." A table summary provides an interpretation or summary of the table's content. For example, "Bermuda's average monthly high temperatures are 68 to 86 degrees Fahrenheit, and the average low temperatures are 60 to 78 degrees Fahrenheit."

Shortcut	Tool Strip	Ribbon
Alt+B, O	Table > Table Properties	Table > Table Properties

To add a caption and summary to a table:

1   Right-click inside a table and select **Table Properties**. The Table Properties dialog box appears.

2   Select the **General** tab.

3   To add a caption:

   □   Type the caption **Text**.

   □   Select a **Side** for the caption.

   □   Select whether the caption should **Repeat** across printed pages.

   □   If the caption repeats, type the text to be appended to the caption on subsequent pages.
        For example: **(cont)**

4   In the **Summary** text box, type a summary.

5   Click **OK**.

Adding alt text to a cross reference or hyperlink

You can add alt text and screen tips to hyperlinks and cross references. The alt text will be read by a screen reader to help users decide if they want to click the link. The screen tip text will appear when the user hovers over the link. It is not normally used by a screen reader, but it can help also users decide if they want to click a link.

Shortcut	Tool Strip	Ribbon
F4 or Crl+K	Insert > Hyperlink	Home > Properties

To add alt text to a cross reference or hyperlink:

1 Select a cross reference or hyperlink and press **F4**.
  —OR—
  Right-click a link and select **Edit Cross Reference** or **Edit Hyperlink**.
  The Insert Cross Reference or Insert Hyperlink dialog box appears.

2 In the **Alternate Text**, type a description of the link's destination.

3 Click **OK**.

Adding alt text to an image

You can add alt text to an image to provide a description for users who can't see the image.

Shortcut	Tool Strip	Ribbon
F4 or Ctrl+G	Insert > Image	Home > Properties

To add alt text to an image:

1 Select an image and press **F4**.
  —OR—
  Right-click an image and select **Image Properties**.
  The Image Properties dialog box appears.

2 Select the **General** tab.

3 In the **Alternate Text** field, type a description of the image.

4 If you want to use the same alt text for all images, select **Apply the alternate text and screen tip to all image references**.

5 Click **OK**.

Adding alt text to a video

You can add alt text to a video to provide a description for users who can't see the video.

Shortcut	Tool Strip	Ribbon
Alt+N, M	Insert > Multimedia	Home > Properties

To add alt text to a video:

1   Select a video and press **F4**.
    —OR—
    Right-click a video and select **Edit Multimedia**.
    The Edit Multimedia dialog box appears.

2   Select the **General** tab.

3   In the **Alternate Text** field, type a description of the video.

4   Click **OK**.

Adding alt text to an equation

You can add alt ("alternate") text to an equation to provide a description of the equation for a screen reader. For example, the equation $E=mc^2$ could use "E equals m c squared" as its alt text.

Shortcut	Tool Strip	Ribbon
Ctrl+E	Insert > Equation	Insert > Equation

To add alt text to an equation:

1   Select an equation and press **Ctrl+E**.
    —OR—
    Right-click an equation and select **Edit Equation**.
    The Equation Editor appears.

2   In the **Alternate Text** field, type the text you want to add.

3   Click **OK**.

Automatically adding alt text

You can automatically add blank alt text to images, QR codes, and equations when you generate a target. Ideally, alt tags should describe the graphical element. However, this is a good option to use to meet accessibility requirements until you have time to write the alt text descriptions.

To automatically add blank alt text images, QR codes, and equations:

1 Open a target.

2 Select the **Advanced** tab.

3 Select **Use empty ALT text for images that do not have ALT text**.

Supporting access to WebHelp when stylesheets are disabled

Section 508 requires content to be accessible without a stylesheet. By default, the WebHelp and WebHelp Plus toolbar and navigation pane don't contain scrollbars. So, some of the content may not be visible, and there will be no way to view it if stylesheets are disabled. You can set your skin to add scrollbars when needed if the user is viewing your content without a stylesheet.

To add scrollbars when stylesheets are disabled:

1 Open a WebHelp skin.

2 Select the **Styles** tab.

3 Open the **Frame** style group.

4 Select the **Toolbar** style.

5 Open the **Frame** property group.

6 Set the **Scrolling** option to **auto**.

Tagging a PDF target

PDF/UA tags allow screen readers and other assistive technologies to identify headings, images, and tables, and other types of content in a document. The tags can then be used to determine the document's reading order and to provide navigational tools.

To include PDF/UA tags:

1 Open a PDF target.

2 Select the **PDF Options** tab.

3 Select **Generate tagged document for PDF/UA**.

Finding and fixing accessibility issues

You can use the Accessibility Suggestions report to find and fix accessibility issues.

To find and fix accessibility issues:

1 Select **Analysis** > **Suggestions** > **Accessibility Suggestions**. The Accessibility Suggestions report appears.

2 To add missing alt text to an image, highlight the image in the list and click **Apply Alt Text**.

3 To fix other issues, double-click the item in the list to open the file that contains the issue.

Additional resources

For more information about accessibility, see:

Wikipedia
en.wikipedia.org/wiki/Web_accessibility

US Government's Section 508 guidelines
www.section508.gov

W3C's Web Accessibility Initiative
www.w3.org/WAI

Internationalization

This section covers:

- Language settings
- Skin translation
- Project translation
- HTML Help localization

Selecting a language for a project

You can select a language for your project to specify the language to be used by the spell checker's dictionary.

Shortcut	Tool Strip	Ribbon
Alt+P, R	(Project toolbar)	Project > Project Properties

To select a language for a project:

1 Select **Project** > **Project Properties**.
   The Project Properties dialog box appears.

2 Select the **Language** tab.

3 Select a language.

4 Click **OK**.

Selecting a language for a topic

You can select a language for a topic to specify the language to be used when spellchecking the topic. A topic's language setting is also used by screen readers and search engines to determine the topic's language.

Shortcut	Tool Strip	Ribbon
F4	Edit > Properties	Home > Properties

To select a language for a topic:

1 Right-click a topic and select **Properties**.
   The Project Properties dialog box appears.

2 Select the **Language** tab.

3 Select a language.

4 Click **OK**.
   A flag will appear at the top of your topic in the XML Editor to indicate the language you selected. Users will not see the flag icon.

Selecting a language for a block of content

You can set the language for a block of content, such as a word, phrase, sentence or paragraph to specify the language to be used by the spell checker for the selected text.

Shortcut	Tool Strip	Ribbon
Alt+H, L, G	Format > Language	Home > Language

To select a language for a block of content:

1 Open a topic.

2 Highlight a block of content.

3 Select **Home** > **Language**.
   The Language dialog box appears.

4 Select a language.

5 Click **OK**.
   A flag will appear before the content in the XML Editor to indicate the language you selected. Users will not see the flag icon.

Finding content with a language tag

When you specify a language for a topic or block of content, Flare adds a language tag to the topic. You can use the Used Language Tags report to find content with an assigned language tag.

To find content with a language tag:

☐ Select **Analysis** > **Used Items** > **Used Language Tags**.
   The Used Language Tags report appears.

Translating a skin

You can modify the default text directly in a skin or in a separate language skin. If you use a language skin, you can reuse your translations with multiple skins.

To translate a skin:

1 Open an HTML5 skin.

2 Select the **UI Text** tab.

3 Select a UI element.

4 Click the **Value** cell and type the translated text.

To create a language skin:

1 Select **File** > **New**.
   The Add File dialog box appears.

2 For **File Type**, select **Language Skin**.

3 Select a **Source** template.

4 Type a **File Name** for the language skin.

5 Click **Add**.
   The language skin is added to the Advanced folder in the
   Project Organizer.

Localizing your content

To localize a project, you (or a translator) will need to translate the
content in your project. The recommended approach is to use MadCap
Lingo to send the project to a translator. Lingo will automatically
include all of the content that needs to be translated. If you try to
collect and send the files yourself, it is extremely likely you forget to
include some of the files, such as your glossary, auto-index phrase set,
or TOC.

The translator can use Lingo or another application to translate the
content. Lingo is integrated with Flare, so it is often the most efficient
approach. However, other translation applications such as SDL Trados
also work well with Flare projects.

Selecting a language for a target

You can select a language in a target to specify the language to use for the labels and tooltips in an online target. For WebHelp targets, the language setting specifies the WebHelp language skin to be used. For HTML5 targets, the language setting specifies the language option from the UI tab of the selected skin.

To select a language for a target:

1 Open a target.

2 Select the **Language** tab.

3 Select a language.

4 If you select a right-to-left language, make sure the right-to-left options are selected.

Including multiple languages in a PDF target

If you translate your project, you can build a PDF that stitches together the PDF targets from multiple projects. For example, you can build your English-language PDF target and set it to also include the PDF target output from a Spanish, French, and German version of the project.

To include multiple languages in a PDF target:

1 Create your project and have it translated.

2 Make sure each language project has a PDF target.

3 Open your primary project.
For example, if you wrote the content in English, the English language project is your primary project.

4 Open your PDF target.

5 Select the **Language** tab.

6 Click [icon].

A new row appears.

7 In the **Linked Flare** project column, click the ... (browse) link.

8 Select a translated project's flprj file and click **Open**.
Flare automatically detects the project's language.

9 Add each language as needed.
[TIP] *The languages will be included in the PDF in order from top to bottom. You can use the arrow buttons to rearrange them.*

10 Build the PDF target.
Each language is included, and Flare adds a heading for each language in the PDF's navigation pane.

Including multiple languages in an HTML5 or WebHelp target

Like PDF targets, HTML5 and WebHelp targets can include the generated output from multiple language projects. For example, you could build an HTML5 target that includes English and Japanese. The users could then click the "Select Skin" button to switch between English and Japanese.

To include multiple languages in an HTML5 or WebHelp target:

1 Create your project and have it translated.

2 Make sure each language project has an HTML5 tripane, HTML top navigation, or WebHelp target.
[icon] *The skin type must match in each project. For example, if you use a topnav skin all of the languages' HTML5 targets should use a topnav skin.*

3 Enable runtime skins in each project. See "Enabling runtime skins" on page 286.

4 Open your primary project.
For example, if you wrote the content in English, the English language project is your primary project.

5 Open your HTML5 target.

6 Select the **Language** tab.

7 Click .
A new row appears.

8 In the Linked Flare project column, click the ... (browse) link.

9 Select a translated project's flprj file and click **Open**.
Flare automatically detects the project's language.

10 Add each language as needed.
TIP *The languages will be listed in the Select Skin toolbar button's drop-down list in order from top to bottom. You can use the arrow buttons to rearrange them.*

11 Build the HTML5 target.

Building localized HTML Help

HTML Help does not provide full support for Unicode, so you may have problems building an HTML Help target that contains Cyrillic or double-byte characters or uses a right-to-left language. One potential fix is to change the system locale setting in Windows before building an HTML Help target, especially if your filenames contain Unicode characters.

To set the Windows system locale:

1 Open the Windows Control Panel.

2 Double-click **Region and Language**.

3 Select the **Administrative** tab.

4 Click **Change System Locale**.

5 Select a language.

6 Click **OK**.

7 Click **Restart Now**.

Flare customization

This section covers customizing Flare's:

- Interface
- Quick Access toolbar
- Quick Launch bar
- Pane layout
- Window layout
- Recent Projects list
- Project analysis data collection
- Auto suggestions
- Macros
- Plug-in API

Interface options

You can turn off Flare's ribbon if you prefer Flare 7 and earlier versions' toolbar-based interface.

To set the interface options:

1   Select **File** > **Flare Options**.
    The Options dialog box appears.

2   Select the **Interface** tab.

3   Select a **Menu Style**.
    If you select **Tool Strip**, Flare's UI will resemble Flare 7.

4   Select a **Theme**.

5   Click **OK**.

Quick Access toolbar options

You can add buttons to the Quick Access toolbar for common tasks. The Quick Access toolbar buttons can be accessed using Alt keyboard shortcuts. For example, the first button (Save) is Alt+1.

To add a button to the Quick Access toolbar:

1   Open the ribbon that contains the button you want to add.

2   Right-click the button and select **Add to Quick Access toolbar**.

To remove a button from the Quick Access toolbar:

☐   Right-click the button and select **Remove from Quick Access toolbar**.

To move the Quick Access toolbar above or below the ribbon:

☐   Right-click the Quick Access toolbar and select **Show Quick Access toolbar below the Ribbon**.

Quick Launch bar

You can use the Quick Launch bar to search for Flare files and commands.

To search for files and commands:

1 Press **Ctrl+Q** or click inside the Quick Launch bar in the top right corner of the Flare window.

2 Type the name of the file or command that you want to find.

3 Click a file or command in the results list.

Pane layout options

By default, the left and right panes organize windows using an accordion. You can change this setting to use tabs.

To use window tabs instead of an accordion:

1 Right-click inside the open pane's title bar.

2 Select **Standard Tabs (Top)** or **Standard Tabs (Bottom)**.

Moving windows

You can open, close, and rearrange Flare's windows or even open a topic or accordion item in a floating window. If you customize the interface, you can save the layout and switch between your layout or the default layout as needed.

To move a window:

1 Click inside the window.

2 Select **Window > Float**.

To reload the default layout:

1 Select **File > Reload Layout**.

2 Click **OK**.

To save the layout:

1 Select **File** > **Save Layout**.

2 Type a name for the layout.

3 Click **OK**.

To select a layout:

1 Select **File** > **Select Layout**.

2 Select a layout.

3 Click **OK**.

Connecting to a proxy server

You can setup Flare to connect to a proxy server, if needed, to login to Salesforce and/or Zendesk.

To connect to a proxy server:

1 Select **File** > **Options**.
The Options dialog box appears.

2 Select the **Project Analysis** tab.

3 Select **Use proxy server**.

4 Type the proxy server's address, and, if required, a port.

5 If your server requires authentication, select **Enable authentication** and type a username and password.

6 Click **OK**.

Pinning a project

You can pin a project in the Recent Projects list on the Start Page. Pinned projects appear at the top of the list.

To pin a project:

1 Open the Start Page.

2 Hover over the recent project in the list.

3 Click ⏎.
The project is pinned to the top of the list.

📖 *You can select **File** > **Manage Recent Projects** to remove projects from the list.*

Project Analysis options

You can turn off the project analysis options to improve Flare's performance. However, it will take longer to generate project analysis reports if the options are turned off.

To set the project analysis options:

1 Select **File** > **Options**.
The Options dialog box appears.

2 Select the **Project Analysis** tab.

3 Select the **Advanced Scan** options.

4 If your project is very large, you can use the **Search Limits** option to reduce the number of search results in project analysis reports.

5 Click **OK**.

Auto suggestion options

You can turn on the auto suggestion options to quickly select and insert frequently used snippets, variables, or other content.

To set the auto suggestion options:

1 Select **File** > **Options**.
The Options dialog box appears.

2 Select whether you want to enable auto suggestions.

3 If you enable auto suggestions:

☐ Select whether you want to enable snippet suggestions.

- ☐ Select the max number of suggestions.

- ☐ Select a minimum character length before the suggestions appear.

- ☐ Select the variable sets, snippets, and/or suggestion term lists to be used to make suggestions.

4 Click **OK**.

To create an auto suggestion list:

1 Select **File** > **New**.
The Add File dialog box appears.

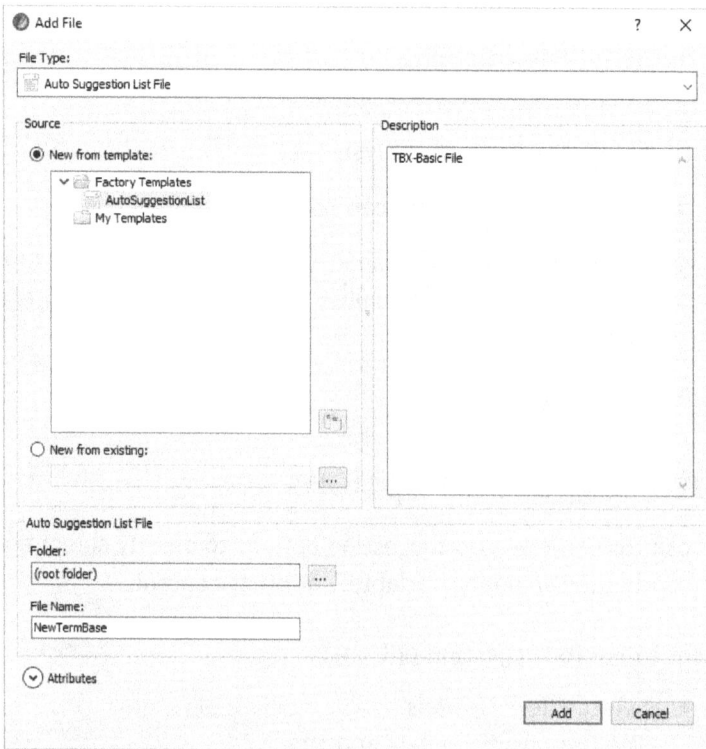

2 For **File Type**, select **Auto Suggestion List File**.

3 Select a **Source Template**.

4 Type a **File Name**.

5 Click **Add**.

The Auto Suggestion File opens in the Auto Suggestion List Editor.

6 Click 📄.

7 Type a phrase.

Macros

You can record macros to automate frequently performed tasks. After you record a macro, you can play the macro by selecting it from the ribbon or the Quick Launch bar. You can also assign a shortcut to the macro or add it to the Quick Access toolbar.

To record a macro:

1 Open a file.

2 Practice performing the task you want to record.

3 Select **Tools** > **Record**.

The New Macro dialog box appears.

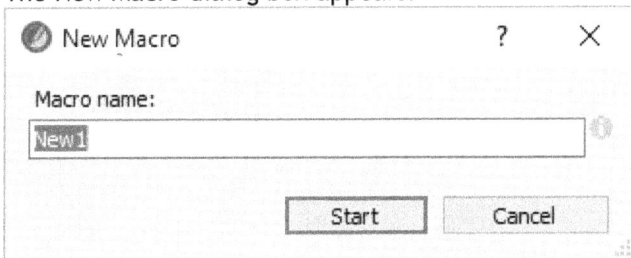

4 Type a **Macro name**.

5 Click **Start**.

6 Perform the steps you want to record.

7 When you are finished, select **Tools** > **Stop**.

To play back a macro:

1 Open a file.

2 Select **Tools** > **Playback**.
A list of macros appears.

3 Select the macro you want to run.

📖 *You can add the "Playback" button to the Quick Access toolbar. If you want to add a keyboard shortcut for a macro, it will appear in the list of commands as "Playback: your macro's name."*

To rename a macro:

1 Open a file.

2 Select **Tools** > **Manage**.
The Manage Macros dialog box appears.

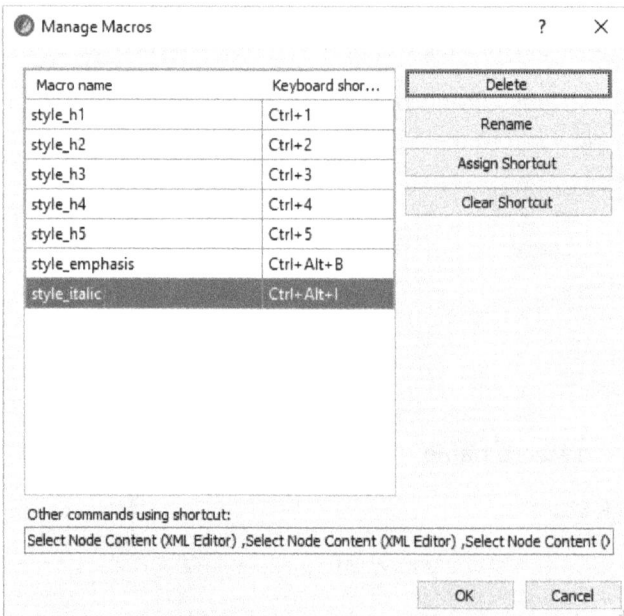

Manage Macros		? ✕
Macro name	**Keyboard shor...**	Delete
style_h1	Ctrl+1	Rename
style_h2	Ctrl+2	
style_h3	Ctrl+3	Assign Shortcut
style_h4	Ctrl+4	Clear Shortcut
style_h5	Ctrl+5	
style_emphasis	Ctrl+Alt+B	
style_italic	Ctrl+Alt+I	

Other commands using shortcut:
Select Node Content (XML Editor) ,Select Node Content (XML Editor) ,Select Node Content ()

OK   Cancel

3 Select a macro.

4 Click **Rename**.

The Rename Macro dialog box appears.

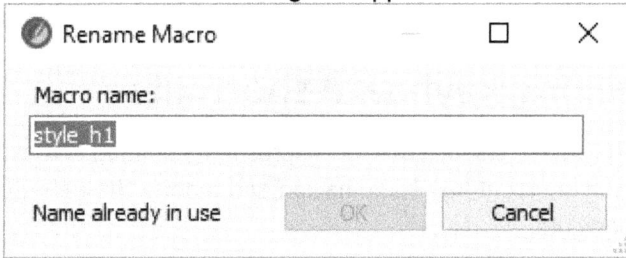

5 Type a new name.

6 Click **OK**.

To delete a macro:

1 Open a file.

2 Select **Tools** > **Manage**.

The Manage Macros dialog box appears.

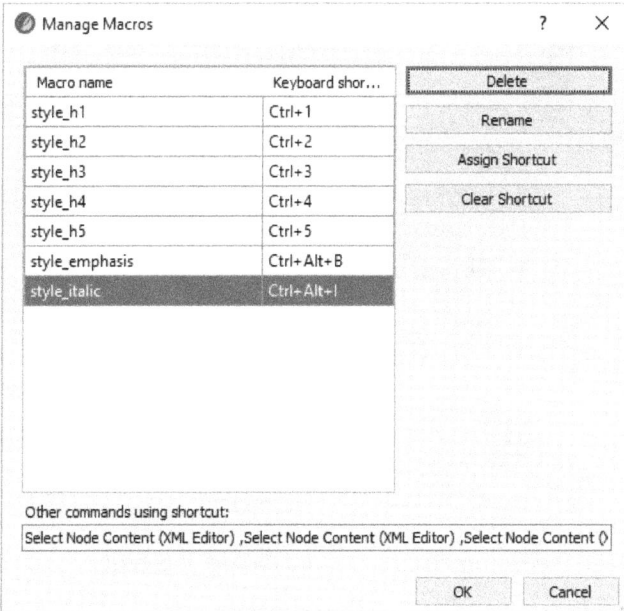

3 Select a macro.

4 Click **Delete**.

5 Click **OK**.

Customizing keyboard shortcuts

You can add keyboard shortcuts for macros and commands that do not have a default shortcut key combination and modify the default assigned shortcuts.

To customize keyboard shortcuts:

1 Select **File** > **Options**.
The Options dialog box appears.

2 Select the **Keyboard Shortcuts** tab.

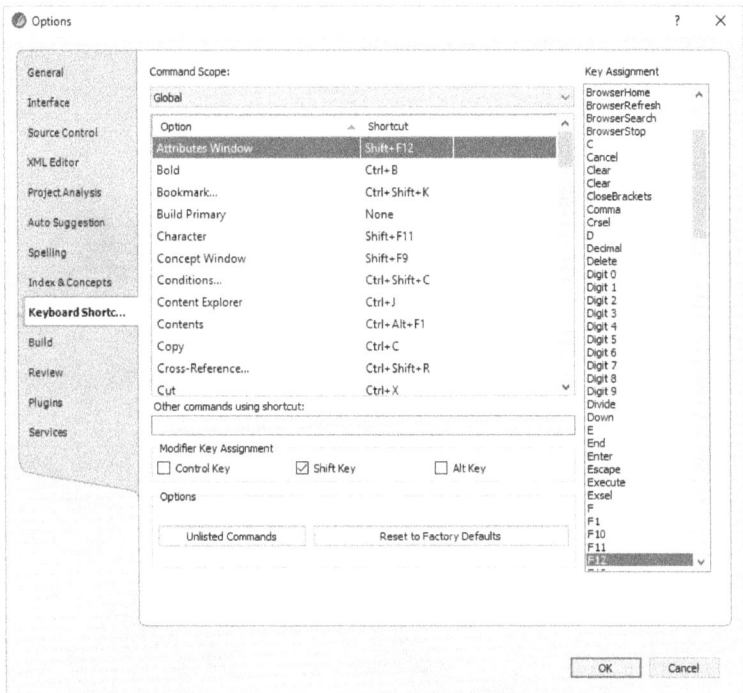

3 Select the **Command** for the keyboard shortcut.

4 For **Key Assignment**, select the key you want to assign.

5 For **Modifier Key Assignment**, select the modifier keys you want to include, such as **Shift**.

6 Click **OK**.

Plugin API

You can use the plugin API to customize or add menus, toolbars, buttons, and features to Flare.

To download the plugin API, see www.madcapsoftware.com/downloads/plugins-api-redistributables.aspx

To add a plugin to Flare:

1 Close Flare.

2 In Windows, open the **Flare.app\Plugins** folder.
By default, the Plugins folder is in C:\Program Files\MadCap Software\MadCap Flare 19\Flare.app.

3 Paste your DLL into the Plugins folder.

4 Open Flare.

5 Select **File** > **Options**.
The Options dialog box appears.

6 Select the **Plugins** tab.

7 Click **Enable**.

8 Click **OK**.

9 Restart Flare.

Downloading Plug-ins

Flare user Mattias Sander has created the excellent and free Kaizen plugin for Flare. You can download it at www.kaizenplugin.com.

MadCap maintains a list of plugins on their website at www.madcapsoftware.com/downloads/plugins-api-redistributables.aspx

Appendices

This section covers:

- Additional resources
- Keyboard shortcuts
- Guide to Flare files
- Quick task index
- Answers to sample questions

Additional resources

Reporting Flare bugs

MadCap does an excellent job of responding to customer problems and fixing bugs. To submit a bug report, select **Help** > **Report a bug**.

Requesting new features

MadCap encourages users to request new features. You can request a feature at www.madcapsoftware.com/feedback/featurerequest.aspx

Crash reporting

If Flare crashes, a dialog box will appear that describes the problem. After you review the information, you can send a crash report.

Flare discussion forums

You can use the Flare discussion forums to research Flare issues, post questions, and meet other Flare users. It's an active and friendly community. You can visit the discussion groups by selecting **Help** > **Help Community** or by visiting www.forums.madcapsoftware.com.

Local help system

Flare's help system is hosted on the web, and MadCap regularly updates it. You can use a local copy of the help if you don't have Internet access while you use Flare.

To use the local help:

1 Select **File** > **Options**.
   The Options dialog box appears.

2 Select the **General** tab.

3 Select **Use Local Help**.

Keyboard shortcuts (by task)

Opening projects, files, and windows

Shortcut	Description
Alt+F4	Close Flare
Ctrl+F4	Close the current window
F4	Open the Properties dialog box from the Content Explorer
Ctrl+Shift+F8	Open the "primary" page layout
Ctrl+Shift+F9	Open the "primary" stylesheet or branding stylesheet **NEW!**
Ctrl+Shift+F12	Open the Attributes window
Alt+A	Open the Analysis ribbon or Table menu
Alt+B	Open the Table ribbon or Build menu
Alt+E	Open the Help ribbon or menu
Alt+F	Open the File menu
Alt+H	Open the Home ribbon or menu
Alt+I	Open the Insert menu
Alt+N	Open the Insert ribbon
Alt+O	Open the Format menu
Alt+P	Open the Project ribbon or menu
Alt+R	Open the Review ribbon or menu
Alt+S	Open the Source Control ribbon or menu
Alt+T	Open the Tools ribbon or menu
Alt+V	Open the View ribbon or menu
Alt+W	Open the Window ribbon or menu

Shortcut	Description
Ctrl+O	Open
Ctrl+T	Create a new file
Ctrl+Tab	Open a "task switcher" popup displaying icons for all open windows. Press Ctrl+Tab again to move through the list.
Ctrl+Shift+Tab	After pressing Ctrl+Tab, move backward through open window list.
Ctrl+Shift+P	Open the Properties dialog box from within the XML Editor

Selecting, moving, cutting, copying, and deleting content

Shortcut	Description
Alt+Shift+↑	Move the selected blocks of content up
Alt+Shift+↓	Move the selected blocks of content down
Ctrl+↑	Move the cursor to the previous block
Ctrl+↓	Move the cursor to the next block
Ctrl+←	Move the cursor to the previous word
Ctrl+→	Move the cursor to the next word
Ctrl+Shift+↑	Select the previous block
Ctrl+Shift+↓	Select the next block
Ctrl+Shift+←	Select the previous word
Ctrl+Shift+→	Select the next word

Shortcut	Description
Ctrl+1	List matches (Auto Suggestion)
Ctrl+2	List frequent phrases (Auto Suggestion)
Ctrl+3	List variables (Auto Suggestion)
Ctrl+A	Select all
Ctrl+C	Copy
Ctrl+V	Paste
Ctrl+X	Cut
Ctrl+Y	Redo
Ctrl+Z	Undo
Ctrl+Backspace	Delete text to the left until the next space
Ctrl+Delete	Delete text to the right to the next space
Ctrl+Insert	Copy
Shift+Delete	Cut
Shift+Insert	Paste
Shift+Tab	Tables: Select the previous cell Lists: Convert to paragraph
Shift+↑	Select the content to the top of the screen (Web layout)
Shift+↓	Select the content to the end of the screen (Web layout)
Del	Delete

Shortcut	Description
End	Move cursor to the end of the line
Home	Move cursor to the beginning of the line
PgUp	Move cursor to previous screen/page
PgDn	Move cursor to next screen/page
Tab	Tables: Select the next cell, or when you press Tab in the last row of a table, add a new row Lists: Indent
Shift+End	Select content between cursor and the end of the line

Formatting and tagging content

Shortcut	Description
F12	Open the Style window
Ctrl+F12	Open the Local Formatting window
Ctrl+Alt+B	Open the Paragraph (or Cell) Properties dialog box
Ctrl+Shift+B	Open the Font Properties dialog box
Ctrl+Shift+C	Open the Condition Tags dialog box
Ctrl+Shift+H	Open the Style Picker
Shift+F12	Open the Attributes window
Ctrl+B	Bold
Ctrl+I	Italic
Ctrl+U	Underline

Inserting content

Shortcut	Description
F11	Insert quick character
Alt+Ctrl+C	Insert ©
Alt+Ctrl+.	Insert ellipsis (...)
Alt+Ctrl+R	Insert ®
Alt+Ctrl+T	Insert ™
Alt+Ctrl+- (numpad)	Insert em dash (—)
Ctrl+;	Insert a paragraph in a list
Ctrl+E	Insert an equation
Ctrl+G	Insert a graphic
Ctrl+R	Insert a snippet
Ctrl+- (numpad)	Insert en dash (–)
Ctrl+Shift+Q	Insert a QR code
Ctrl+Shift+V	Insert a variable
Ctrl+Shift+-	Insert non-breaking hyphen
Shift+Space	Insert non-breaking space
Shift+F11	Open the Character dialog box

Linking

Shortcut	Description
Ctrl+K	Insert a hyperlink
Ctrl+Shift+K	Insert a bookmark

Shortcut	Description
Ctrl+Shift+R	Insert a cross reference
Ctrl+Shift+T	Insert a glossary term link
Shift+F9	Open the Concepts window

Saving

Shortcut	Description
Ctrl+S	Save
Alt+Ctrl+S	Save as
Ctrl+Shift+S	Save all

Spell checking and tracking changes

Shortcut	Description
F7	Open the Spell Check window
Ctrl+Shift+E	Enable track changes

Working with the index and TOC

Shortcut	Description
F2	Highlight TOC entry for editing
F9	Open the Index window
F10	Insert the selected text as an index keyword

Ctrl+F8	Open the "primary" TOC

Searching

Shortcut	Description
F3	Find next
Ctrl+Shift+F	Open the Find and Replace in Files window
Ctrl+F	Open Quick Find widget
Ctrl+H	Open Quick Replace widget
Ctrl+N	Open the Find Elements window

Opening and docking windows

Shortcut	Description
Ctrl+[	Open the Project Organizer
Ctrl+J	Open the Content Explorer
Ctrl+W	Open the preview
Ctrl+`	Toggle XML Editor and Text Editor
Ctrl+Shift+D	Move the active document to the Document Dock
Ctrl+Shift+I	Open the Instant Messages window
Ctrl+Shift+J	Open the File List
Ctrl+Shift+O	Open the Messages window

Shortcut	Description
Ctrl+Shift+W	Open the Start page

Opening Flare's help system

Shortcut	Description
F1	Open a context-sensitive help topic
Alt+F1	Open the help system's index
Alt+Ctrl+F1	Open the help system's TOC
Alt+Shift+F2	Open the help system's index results window
Ctrl+F1	Open the help system's search
Ctrl+F3	Open a context-sensitive help topic in the dynamic help window

Publishing

Shortcut	Description
F6	Build the primary target
Ctrl+F6	Publish the primary target
Ctrl+F9	Open the primary target in Target Editor
Shift+F6	Open the primary target

Keyboard shortcuts (by key)

Shortcut	Description
Alt+Ctrl+B	Open the Paragraph (or Cell) Properties dialog box
Alt+Ctrl+C	Insert ©
Alt+Ctrl+R	Insert ®
Alt+Ctrl+S	Save as
Alt+Ctrl+T	Insert ™
Alt+Ctrl+.	Insert ellipsis (...)
Alt+Ctrl+- (numpad)	Insert em dash (—)
Alt+Ctrl+F1	Open the Flare help system's TOC
Alt+Shift+F2	Open the Flare help system's index results window
Alt+Shift+↑	Move the selected blocks of content up
Alt+Shift+↓	Move the selected blocks of content down
Alt+F1	Open the Flare help system's index
Alt+F4	Close Flare
Alt+1 through 9	Quick Access toolbar buttons
Alt+A	Open the Analysis ribbon or menu
Alt+B	Open the Table ribbon or Build menu
Alt+E	Open the Help ribbon or menu
Alt+F	Open the Find menu
Alt+H	Open the Home ribbon or menu
Alt+I	Open the Insert menu
Alt+N	Open the Insert ribbon
Alt+P	Open the Project ribbon or menu
Alt+R	Open the Review ribbon or menu
Alt+S	Open the Source Control ribbon or menu

Shortcut	Description
Alt+T	Open the Tools ribbon or menu
Alt+V	Open the View ribbon or menu
Alt+W	Open the Window ribbon or menu
Alt+F4	Close Flare
Ctrl+F1	Open the Flare help system's search
Ctrl+F2	Open file in XML Editor
Ctrl+F3	Open a context-sensitive help topic
Ctrl+F4	Close active window
Ctrl+F6	Publish the primary target
Ctrl+F8	Open the "primary" TOC
Ctrl+F9	Open Primary target in Target Editor
Ctrl+F12	Open the Formatting window
Ctrl+1	List matches (Auto Suggestion)
Ctrl+2	List frequent phrases (Auto Suggestion)
Ctrl+3	List variables (Auto Suggestion)
Ctrl+A	Select all
Ctrl+B	Bold
Ctrl+C	Copy
Ctrl+E	Insert an equation
Ctrl+F	Open the Quick Find widget
Ctrl+G	Insert a graphic
Ctrl+H	Open the Quick Replace widget
Ctrl+I	Italic
Ctrl+J	Open the Content Explorer
Ctrl+K	Insert a hyperlink
Ctrl+N	Open the Find Elements window
Ctrl+O	Open

Shortcut	Description
Ctrl+P	Print
Ctrl+Q	Move the insertion point to the Quick Launch bar
Ctrl+R	Insert a snippet
Ctrl+S	Save
Ctrl+T	Create a new file
Ctrl+U	Underline
Ctrl+V	Paste
Ctrl+W	Open the preview
Ctrl+X	Cut
Ctrl+Y	Redo
Ctrl+Z	Undo
Ctrl+0	Zoom 100% (XML Editor) / reset text size (Text Editor)
Ctrl++	Zoom In in XML Editor
Ctrl+-	Zoom Out in XML Editor
Ctrl+-	Decrease text size in Text Editor
Ctrl+=	Increase text size in Text Editor
Ctrl+`	Toggle XML Editor and Text Editor
Ctrl+[	Open the Project Organizer
Ctrl+;	Insert a paragraph in a list
Ctrl+Backspace	Delete text to the left until the next space
Ctrl+Delete	Delete text to the right to the next space
Ctrl+Insert	Copy
Ctrl+Tab	Tab between UI elements
Ctrl+↑	Move to the beginning of the previous block
Ctrl+↓	Move the beginning of the next block
Ctrl+←	Move to the beginning of the previous word
Ctrl+→	Move to the beginning of the next word

Shortcut	Description
Ctrl+- (numpad)	Insert en dash (–)
Ctrl+Shift+F8	Open the "primary" page layout
Ctrl+Shift+F9	Open the "primary" stylesheet or branding stylesheet **NEW!**
Ctrl+Shift+F12	Open the Attributes window
Ctrl+Shift+B	Open the Font Properties dialog box
Ctrl+Shift+C	Open the Condition Tags dialog box
Ctrl+Shift+D	Move the active document to the Document Dock
Ctrl+Shift+E	Enable track changes
Ctrl+Shift+F	Open the Find and Replace in Files window
Ctrl+Shift+H	Open the Style Picker
Ctrl+Shift+I	Open the Instant Messages window
Ctrl+Shift+J	Open the File List
Ctrl+Shift+K	Insert a bookmark
Ctrl+Shift+O	Open the Messages window
Ctrl+Shift+Q	Insert QR Code
Ctrl+Shift+P	Open the Properties dialog box from within the XML Editor
Ctrl+Shift+R	Insert a cross reference
Ctrl+Shift+S	Save all
Ctrl+Shift+T	Insert a glossary term link
Ctrl+Shift+V	Insert a variable
Ctrl+Shift+W	Open the Start page
Ctrl+Shift +-	Insert non-breaking hyphen
Ctrl+Shift +↑	Select content to beginning of content block
Ctrl+Shift +↓	Select content to end of content block
Ctrl+Shift+←	Select previous word
Ctrl+Shift+→	Select next word

Shortcut	Description
Ctrl+Shift+Tab	After pressing Ctrl+Tab, move backward through open window list
F1	Open a context-sensitive help topic
F2	Highlight for editing
F3	Find next
F4	Open the Properties dialog box from the Content Explorer
F5	Refresh
F6	Build the primary target
F7	Open the Spell Check window
F8	Repeat last action in XML Editor
F9	Open the Index window
F10	Insert the selected text as an index keyword
F11	Insert quick character
F12	Open the Style window
Shift+F6	Open the primary target
Shift+F7	Open the Thesaurus window
Shift+F9	Open the Concepts window
Shift+F11	Open the Character dialog box
Shift+F12	Open the Attributes window
Shift+←	Select the previous character
Shift+→	Select the next character
Shift+Delete	Cut
Shift+End	Select content between cursor and the end of the line
Shift+Home	Select content between cursor and beginning of the line
Shift+Insert	Paste
Shift+Space	Insert non-breaking space
Shift+Tab	Tables: Select the previous cell

Shortcut	Description
	Lists: Convert to paragraph
Shift+PgUp	Select content to end of screen (Web layout)
Shift+PgDn	Select content to top of screen (Web layout)
Del	Delete
Tab	Tables: Select the next cell, or, if you press Tab in the last row of a table, add a new row
	Lists: Indent
End	Move to the end of the line
PgUp	Move to previous screen/page
PgDn	Move to next screen/page

Guide to Flare files

The following table lists the file types that are used in Flare, their extension, and their default folder.

File Type	Extension	Default Folder
Analyzer database	fldb	Analyzer\Content.cadbf
Auto suggestion list	fltbx	Project\Advanced
Batch target	flbat	Project\Targets
Branding Stylesheet **NEW!**	css	Content\Resources\Branding
Browse sequence	flbrs	Project\Advanced
Confluence import file	flimpconf	Project\Imports
Condition tag set	flcts	Project\ConditionTagSets
Context-sensitive help alias file	flali	Project\Advanced
Context-sensitive help header file	h or hh	Project\Advanced
Contribution template	mccot	Documents\My Contribution Templates
Dependency set	fllnks	Output\Temporary
Dictionary (Sentry, old format)	clx, tlx	Program Files\MadCap Software\ MadCap Flare 19\Flare.app\Resources\ SSCE

File Type	Extension	Default Folder
Dictionary (Hunspell)	oxt	Program Files\MadCap Software\ MadCap Flare 19\Flare.app\Resources\ HunspellDict
DITA topic	dita	Output\name of DITA target
DITA map	ditamap	Output\name of DITA target
DITA import file	flimpdita	Project\Imports
EPUB file	epub	Output\name of EPUB target
Excel import file	flimpxls	Project\Imports
Flare ZIP file	flprjzip	Documents
Fontset	mcfst	AppData\Roaming\MadCap Software\Flare
FrameMaker import file	flimpfm	Project\Imports
Git ignore	.ignore	Project's folder
Glossary	flglo	Project\Glossaries
HTML Help	chm	Output\name of HTML Help target
Image	bmp, emf, eps, exps, gif, hdp, jpg, png, ps, svg, tif, webm, wdp, wmf, xaml, xps	Content\Resources\Images
Index auto-index phrase set	flaix	Project\Advanced
Index link set	flixl	Project\Advanced
Keyboard shortcuts	mccmds	AppData\Roaming\MadCap Software\Flare
Language skin	fllng	Program Files\MadCap Software\ MadCap Flare 19\Flare.app\Resources\ LanguageSkins

File Type	Extension	Default Folder
Markdown import file	flimpmd	Project\Imports
Meta tag set	flmeta	Project\Advanced
Macros	none	AppData\Roaming\MadCap Software\Flare\Macros
Micro content	flmco	Content\Resources\MicroContent
Page layout	flpgl	Content\Resources\PageLayouts
PDF file	pdf	Output\name of PDF target
Project file	flprj	top-level folder
Project import file	flimpfl	Project\Imports
Publishing destination	fldes	Project\Destinations
Relationship table	flrtb	Project\Advanced
Report	flrep	Project\Reports
Review package	fltrev	Content
Search filter set	flsfs	Project\Advanced
Search synonyms	mcsyns	Project\Advanced
Skin	flskn	Project\Skins
Snippet	flsnp	Content\Resources\Snippets
Sound	au, midi, mp3, opus, wav, wma	Content\Resources\Multimedia
Stylesheet	css	Content\Resources\Stylesheets
Table stylesheet	css	Content\Resources\TableStyles
Target	fltar	Project\Targets

File Type	Extension	Default Folder
Target build error log	mclog	Project\Reports
Template	htm	Documents\My Templates\Content
Template page	flmsp	Content\Resources\TemplatePages
TOC	fltoc	Project\TOCs
Topic (generated)	htm	Output
Topic (source)	htm	Content
Variable set	flvar	Project\VariableSets
Video	asf, avi, mov, mp4, mpg, ogg, ogv, qt, swf, u3d, webm, wmv	Content\Resources\Multimedia
Window layout	panellayout	AppData\Roaming\MadCap Software\Flare
Word import file	flimp	Project\Imports
XHTML document	xhtml	Output\name of XHTML target

Quick task index

The quick task index provides the basic steps for every major task you can perform in Flare.

Concept	See Page
Projects	516
Topics	519
Topic content	522
Tables	524
Links	524
Navigational tools	528
Formatting and design	530
Variables and snippets	531
Condition tags	533
Targets	534
Context-sensitive help	535
Templates	536
File tags and reports	537
Annotations and topic reviews	537
Pulse	539
Source control	540
eLearning	543

Projects

Task	Steps	See Page
Creating a project	1 Select **File** > **New Project** > **New Project**. 2 Type a **Project Name** and **Project Folder**. 3 Select a **Language** and click **Next**. 4 Select a **Template Folder** and **Template** and click **Next**. 5 If needed, customize the design elements for your project's branding stylesheet. 6 Select an **Available Target** and click **Finish**.	33
Importing an Author-it project	1 Publish your Author-it project as XML. 2 Download and install the MadCap Author-it to Flare Converter. 3 Open the MadCap Author-it to Flare Converter. 4 Locate and select your published XML file. 5 Select a **Project Folder**. 6 Select a **Content Path Option**. 7 If your Author-it project contains multiple books, select a **Book**. 8 Click **Convert**.	35
Creating a project based on Confluence pages	1 Select **File** > **New Project** > **Confluence Pages**. 2 Type a **Project Name**. 3 Type or select a **Project Folder**. 4 Select an **Output Type**. 5 Type the path to the **Confluence Server**. 6 Type your **Username**. 7 Type your **Password/API Token**. 8 Click **Submit**. 9 Select a **Space** and the pages you want to import. 10 Click the **Advanced Options** tab. 11 Select whether you want to **Import linked pages, Remove inline formatting, Remove style classes,**and/or **Import resources**. 12 Click **Finish**.	37

Task	Steps	See Page
Importing a Doc-to-Help project	1 Select **File** > **New Project** > **Doc-to-Help Project**. 2 Select a project and click **Open**. 3 Click **Next**. 4 Type a **Project Name**, select a **Project Folder**, and click **Next**. 5 Select the folder where you store your stylesheet(s), click **Select Folder**, and click **Next**. 6 Select a language for the spell checker and click **Finish**.	35
Importing a RoboHelp project	1 Select **File** > **New Project** > **RoboHelp Project**. 2 Select a project and click **Open**. 3 Click **Next**. 4 Type a **Project Name**, select a **Project Folder**, and click **Next**. 5 Select whether you want to **Convert all topics at once** and/or to **Convert inline formatting to CSS styles** and click **Next**. 6 Select a language for the spell checker and click **Finish**.	56
Importing an HTML Help file	1 Select **File** > **New Project** > **HTML Help Project (HHP)**. 2 Select a project and click **Open**. 3 Click **Next**. 4 Type a **Project Name**, select a **Project Folder**, and click **Next**. 5 Select whether you want to **Convert all topics at once** and/or to **Convert inline formatting to CSS styles** and click **Next**. 6 Select a language for the spell checker and click **Finish**.	52
Creating a project based on a FrameMaker document	1 Select **File** > **New Project** > **FrameMaker Documents**. 2 Type a **Project Name** and type or select a **Project Folder**. 3 Select an **Output Type** and click **Next**.	57

Task	Steps	See Page
	4 Click ⊞, select an FM or book file, and click **Open**.	
	5 Select whether you want to link to the original FrameMaker document and click **Next**.	
	6 Select a style or styles to use to create new topics and click **Next**.	
	7 Select how you want to import images and whether you want to import table styles.	
	8 Select whether you want to automatically reimport from FrameMaker whenever you build a target and click **Next**.	
	9 Select a stylesheet and click **Next**.	
	10 Map your styles and click **Finish**.	
Creating a project based on a Word document	1 Select **File** > **New Project** > **Word Documents**.	83
	2 Type a **Project Name** and type or select a **Project Folder**.	
	3 Select an **Output Type** and click **Next**.	
	4 Click **Add File**.	
	5 Select a Word document and click **Open**.	
	6 Select the **Styles** tab.	
	7 Select a stylesheet for the new topic(s).	
	8 Select a style or styles to use to create new topics.	
	9 Map your paragraph and character styles.	
	10 Select the styles to use to create new topics.	
	11 Select the **Advanced Options** tab.	
	12 Select **Link generated files to source files** if you plan to continue editing in Word.	
	13 Click **Import** and click **Accept**.	
Creating a project based on a DITA document set	1 Select **File** > **New Project** > **DITA Document Set**.	95
	2 Type a **Project Name** and type or select a **Project Folder**.	
	3 Select an **Output Type** and click **Next**.	
	4 Click ⊞, select a .dita or .ditamap file, and click **Open**.	

Task	Steps	See Page
	5 If you plan to continue editing the original DITA files, select **Link generated files to source files.**	
	6 Click **Next.**	
	7 Type a **Project Name.**	
	8 Type or select a **Project Folder** and click **Next.**	
	9 Select **Import all content files to one folder** if you want to import all of the DITA documents into one folder.	
	10 Select '**Auto-reimport before Generate Output'** if you want to automatically re-import the DITA document(s) when you generate a target.	
	11 Select **Preserve ID attributes for elements** if you plan to build a DITA target from your project.	
	12 Click **Next.**	
	13 Click **Conversion Styles** if you want to change the formatting of your topics.	
	14 Select a stylesheet for the new topic(s).	
	15 Click **Finish.**	

Topics

Task	Steps	See Page
Creating a topic	1 Select **File > New.**	69
	2 For **File Type**, select **Topic.**	
	3 Select a **Source** template.	
	4 Type a **File Name** and click **Add.**	
Importing a Word document	1 Create or open an MS Word import file.	83
	2 Click **Add File.**	
	3 Select a Word document and click **Open.**	
	4 Select the **Styles** tab.	
	5 Select a stylesheet for the new topic(s).	
	6 Select style(s) to use to create new topics.	

Task	Steps	See Page
	7 Map your paragraph and character styles.	
	8 Select the styles to use to create new topics.	
	9 Select the **Advanced Options** tab.	
	10 Select **Link generated files to source files** if you plan to continue editing in Word.	
	11 Click **Import** and click **Accept**.	
Importing a FrameMaker document	1 Create or open a FrameMaker import file.	56
	2 Click ⊕.	
	3 Select a FrameMaker .fm or .book document and click **Open**.	
	4 Select whether you want to link the generated files to the source files.	
	5 Select the **New Topic Styles** tab and select the styles to use to create new topics.	
	6 Select the **Options** tab and select how you want to import image and table styles.	
	7 Select the **Stylesheet** tab and select a stylesheet.	
	8 Select the **Paragraph Styles** tab and map your paragraph styles.	
	9 Select the **Character Styles** tab and map your character styles.	
	10 Select the **Cross Reference Styles** tab and map your cross-reference (x-ref) styles.	
	11 Click **Import** and click **Accept**.	
Importing Confluence pages	1 Create or open a Confluence Import file.	39
	2 Select or type a folder for the imported topics.	
	3 Type the path to the **Confluence Server**.	
	4 Type your **Username** and **Password/API Token**.	
	5 Select the **Workspace Selection** tab and select a **Space** and the pages you want to import.	
	6 Select the **Advanced Options** tab.	
	7 Select whether you want to **Import linked pages**, **Remove inline formatting**, **Remove style classes**, and/or **Import resources**.	

Task	Steps	See Page
	8 Select **Link generated files to source files** if you plan to continue editing the original Confluence pages.	
	9 Click **Import** and click **Accept**.	
Importing a DITA file	1 Create or open a DITA import file.	89
	2 Click ⊕.	
	3 Select a .dita or .ditamap file and click **Open**.	
	4 Select whether you want to link the generated files to the source files.	
	5 Select the **Options** tab and select whether you want to import your content into one folder.	
	6 Select the **Stylesheet** tab and select a stylesheet.	
	7 Click **Import** and click **Accept**.	
Importing an HTML file	1 Create or open an HTML import file.	100
	2 Click ⊕.	
	3 Select a .htm, .html, or .xhtml document and click **Open**.	
	4 Select whether you want to link the generated files to the source files.	
	5 Select a folder for the imported topics.	
	6 Select **Import resources** if you also want to import any files that are used by the selected document(s).	
	7 Click **Import** and click **Accept**.	
Importing a Markdown file	1 Create or open a Markdown import file.	102
	2 Click **Add File**.	
	3 Select a Markdown file and click Open.	
	4 Select the **Styles** tab.	
	5 Select a stylesheet for the new topic(s).	
	6 Select style(s) to use to create new topics.	
	7 Map your Markdown styles to Flare styles.	
	8 Select the styles to use to create new topics.	
	9 Select the **Advanced Options** tab.	
	10 Select **Link generated files to source files** if you plan to continue editing the Markdown files.	

Task	Steps	See Page
	11 Click **Import** and click **Accept**.	
Importing an external resource	1 Select **Project > External Resources**.	105
	2 Click 📇.	
	3 Select the folder that contains the external resource.	
	4 Click **OK**.	
	5 Select the file(s) you want to import.	
	6 Click 📋.	
	7 Select a folder and click **OK**.	
Importing content from Flare projects	1 Create or open a Flare project import file.	100
	2 Click **Browse**.	
	3 Select a Flare project file and click **Open**.	
	4 Select whether you want to automatically re-import the files when you build a target.	
	5 For **Include Files**, select the files or file types to be imported.	
	6 For **Exclude Files**, select the file types to not be imported.	
	7 Click **Import**.	

Topic content

Task	Steps	See Page
Expanding/ collapsing headings	1 Position the cursor to the left of a heading.	74
	2 Click the triangular expand/collapse icon.	
Inserting a special character	1 Select **Insert > Character**.	74
	2 Select a character.	
Inserting a QR code	1 Select **Insert > QR Code**.	76
	2 Select a **Content Type**.	
	3 Type the **Content**.	
	4 Select a **Size**.	

Task	Steps	See Page
Inserting a code snippet	1 Select **Insert** > **Code Snippet**. 2 Type or paste the code snippet. 3 Select a **Language**.	78
Inserting an equation	1 Select **Insert** > **Equation**. 2 Use the ribbons and toolbars to create the equation.	77
Inserting an iframe	1 Select **Insert** > **IFrame**. 2 Type a **Source** URL. 3 Select or type a **Width**. 4 Select or type a **Height**. 5 Click **OK**.	79
Creating a list	1 Click the down arrow to the right of the ⁝≡ ˅ button in the Home ribbon. 2 Select a list type. 3 Type the list items.	111
Sorting a list	1 Right-click the list's ol or ul tag. 2 Select **Sort List**.	112
Inserting an image	1 Select **Insert** > **Image**. 2 Click **Browse**, select an image, and click **Open**. 3 Type an **Alternate Text** description. 4 Click **OK**.	129
Inserting a PDF image	1 Select **Insert** > **Image**. 2 Click **Browse**, select a PDF, and click **Open**. 3 Select a **Page**. 4 Type an **Alternate Text** description. 5 Click **OK**.	131
Inserting a sound or video	1 Select **Insert** > **Multimedia** > *your movie type*. 2 Click **Browse**, select a sound or video file, and click **Open**. 3 Type an **Alternate Text** description. 4 Click **OK**.	131, 134

Task	Steps	See Page
Inserting a YouTube or Vimeo video	1 Select **Insert** > **Multimedia** > **YouTube/Vimeo**. 2 Select **Multimedia from Web**. 3 Type or paste a URL. 4 Type an **Alternate Text** description. 5 Click **OK**.	134
Inserting a 3D model	1 Select **Insert** > **Multimedia** > **3D Model**. 2 Click **Browse**, select a 3D model, and click **Open**. 3 Type an **Alternate Text** description. 4 Click **OK**.	135

Tables

Task	Steps	See Page
Inserting a table	1 Select **Insert** > **Table**. 2 Select a number of columns and rows. 3 Select a number of header and footer rows. 4 Type a table caption and select a location. 5 Type a table summary. 6 Select a column width. 7 Click **OK**.	114
Creating a table style	1 Select **File** > **New**. 2 For **File Type**, select **Table Style**. 3 Select a **Source** template. 4 Select a **Folder**. 5 Type a **File Name**. 6 Click **Add**.	116
Converting text to a table	1 Highlight the text. 2 Select **Insert** > **Table**. 3 In the **Text to Table** group box, select a conversion option. 4 Click **OK**.	116

Task	Steps	See Page
Converting a table to text	1 Select the table. 2 Select **Table > Convert to Text.**	116
Sorting table rows	1 Click inside the column you want to use for sorting. 2 Select **Table > Sort Rows > Ascending** or **Descending.**	116
Rearranging table rows or columns	1 Select a row in the tag bar or a column in the span bar. 2 Drag the row up/down or the column left/right. 3 Release the mouse button to move the row or column.	117
Assigning a table style to a table	1 Click inside a table. 2 Click **Apply Table Style.** 3 Select a **Table Style.**	125
Assigning a table style to multiple tables	1 Open a table style. 2 Click **Apply Style.** 3 Select a topic or folder. 4 Click **OK.**	126
Removing inline formatting from a table	☐ Select **Table > Reset Local Cell Formatting.**	127

Links

Task	Steps	See Page
Creating a hyperlink	1 Select the text or image to use as the link. 2 Select **Insert > Hyperlink.** 3 Select a link target (a topic, file, or website). 4 Select a **Target Frame.** 5 Type the **Alternate Text.** 6 Click **OK.**	149

Task	Steps	See Page
Creating an image map link	1 Right-click an image and select **Image Map**. 2 Select an image map shape and draw the image map area. 3 Select a link target type and target. 4 Select a **Target Frame**. 5 Type the **Alternate Text**. 6 Click **OK**.	156
Creating a topic popup	1 Select the text or image to use as the link. 2 Select **Insert** > **Hyperlink** > **Topic Popup**. 3 Select a link target (a topic, file, or website). 4 Type the **Alternate Text**. 5 Click **OK**.	159
Creating a text popup	1 Select the text or image that you want to use as the link. 2 Select **Insert** > **Text Popup**. 3 Type the popup text. 4 Click **OK**.	160
Creating a cross reference	1 Position your cursor where you want to add the cross reference. 2 Select **Insert** > **Cross Reference**. 3 For **Link To**, select **Topic in Project**. 4 Type the **Alternate Text**. 5 Select a topic and click **OK**.	162
Finding and fixing broken links	1 Select **Analysis** > **Links** > **Broken Links**. 2 Double-click a broken link in the list. 3 Right-click the highlighted link and select **Edit Hyperlink**. 4 Select a new link location. 5 Click **OK**.	157

Drop-down, expanding, and toggler links

Task	Steps	See Page
Creating a drop-down link	1 Highlight the drop-down link.	165
	2 Select **Insert > Drop-Down Text**.	
	3 Type or cut and paste the drop-down content into the blue bracket.	
	4 Click **OK**.	
Creating an expanding link	1 Highlight the expanding link.	166
	2 Select **Insert > Expanding Text**.	
	3 Type or cut and paste the expanding text into the blue bracket.	
	4 Click **OK**.	
Creating a toggler link	1 Right-click the tag bar next to the content you want to show and hide.	166
	2 In the popup menu, select **Name**.	
	3 Type a name and click **OK**.	
	4 Highlight the toggler link text.	
	5 Select **Insert > Toggler**.	
	6 Select a toggler target.	
	7 Click **OK**.	

Related topic, keyword, and concept links

Task	Steps	See Page
Creating a related topics link	1 Select **Insert > Related Topics Control**.	169
	2 Select a topic to add to the link.	
	3 Click ➡ to add the topic to the related topics link.	
	4 Add more topics as needed.	
	5 Click **OK**.	
Creating a keyword link	1 Select **Insert > Keyword Link Control**.	169
	2 Select a keyword.	
	3 Click ➡ to add the keyword to the keyword link.	

Task	Steps	See Page
	4 Add more keywords as needed.	
	5 Click **OK**.	
Creating a concept link	1 Add a concept term to a topic or topics.	170
	2 Select **Insert** > **Concept Link**.	
	3 Select a concept.	
	4 Click ➡ to add the concept to the concept link.	
	5 Click **OK**.	

Relationship links

Task	Steps	See Page
Creating a relationship table	1 Select **File** > **New**.	174
	2 For **File Type**, select **Relationship Table**.	
	3 Select a **Template Folder** and **Template**.	
	4 Type a **File Name**.	
	5 Click **Add**.	
Adding a relationship to a relationship table	1 Open a relationship table.	175
	2 Click 🗂 to create a new row.	
	3 Click 🗐.	
	4 Type a name for the row.	
	5 Click **OK**.	
	6 Click a cell.	
	7 Click 📄.	
	8 Select a topic and click **OK**.	
Creating a relationship link	1 Select **Insert** > **Proxy** > **Relationships Proxy**.	177
	2 Click **OK**.	

Navigational tools

Task	Steps	See Page
Creating a TOC book	1 Click ▣. 2 Press **F2**. 3 Type a name.	188
Creating a TOC page	1 Click ▣. 2 Double-click the TOC page. 3 Type a **Label** for the page. 4 Click **Select Link**. 5 Select a topic. 6 Click **Open**. 7 Click **OK**.	188
Finding and fixing issues in a TOC	1 If your TOC books are intentionally unlinked, click ▣. 2 Click ▣. 3 Right-click the selected TOC item and select **Properties**. 4 On the **General** tab, select a new link and click **OK**.	189
Finding topics that are not in a TOC	1 Select **Analysis > More Reports > Topics Not In Selected TOC**. 2 For **Filter**, select a TOC.	195
Creating an index entry	1 Position the cursor where you want to insert the index entry marker. 2 Press **F9**. 3 Type the index entry and press **Enter**.	198
Finding topics that are not in the index	1 Select **Analysis > Topics Not In Index**. 2 Double-click a topic in the list to open it and add keywords.	204
Excluding a topic from the search	1 Right-click a topic and select **Properties**. 2 Select the **Topic Properties** tab. 3 Deselect the **Include topic when full-text search database is generated** option. 4 Click **OK**.	210

Task	Steps	See Page
Creating micro content	1 Select **File** > **New**. 2 For **File Type**, select **Micro Content**. 3 Select a **Source** template. 4 Type a **File Name**. 5 Click **Add**. 6 Click . 7 Type a search phrase and press **Enter**. 8 Add a response.	324
Creating micro content from existing content	1 Highlight the content you want to convert to a micro content. 2 Select **Home** > **Create Micro Content**. 3 Type a **Phrase**. 4 Select a **Micro Content File**. 5 Click **OK**.	324
Creating a glossary entry	1 Click . 2 Type a **Term**. 3 Type a **Definition** or select a topic for the definition. 4 Click **OK**.	216
Creating a browse sequence	1 Select **File** > **New**. 2 For **File Type**, select **Browse Sequence**. 3 Select a **Source** template. 4 Type a **File Name**. 5 Click **Add**.	222

Formatting and design

Task	Steps	See Page
Creating a stylesheet	1 Select **File** > **New**. 2 For **File Type**, select **Stylesheet**. 3 Select a **Source** template. 4 Type a **File Name**.	234

Task	Steps	See Page
	5 Click **Add**.	
Creating a style	1 Use the **Home** ribbon commands to format the content.	237
	2 Select the formatted content.	
	3 Select **Home** > **Style Window**.	
	4 Click **Create Style**.	
	5 Type a **Class** name for the new style.	
	6 Click **OK**.	
Creating a branding stylesheet **NEW!**	1 Select **File** > **New**.	260
	2 For **File Type**, select **Branding**.	
	3 Select a **Source** template.	
	4 Type a **File Name**.	
	5 Click **Add**.	
Creating a page layout	1 Select **File** > **New**.	264
	2 For **File Type**, select **Page Layout**.	
	3 Select a **Source** template.	
	4 Type a **File Name**.	
	5 Click **Add**.	
Creating a template page	1 Select **File** > **New**.	272
	2 For **File Type**, select **Template Page**.	
	3 Select a **Source** template.	
	4 Type a **File Name**.	
	5 Click **Add**.	
Creating a skin	1 Select **File** > **New**.	276
	2 For **File Type**, select **Skin**.	
	3 Select a **Source** template.	
	4 Type a **File Name**.	
	5 Click **Add**.	
	6 Modify the skin options as needed.	
Creating a responsive layout	1 Position the cursor where you want to insert the responsive layout.	288
	2 Select **Home** > **Responsive Layout**.	
	3 Click **New Style**.	
	4 Type a **Class Name**.	

Task	Steps		See Page
	5	Select a **Stylesheet** and **Row Template** and click **OK**.	
	6	Click **Insert Row**.	
	7	Click **Web**, click inside each cell, and select a **Cell Width** and **Cell Offset**.	
	8	Click **Tablet** and set the cell widths and offsets.	
	9	Click **Mobile** and set the cell widths and offsets.	

Variables and snippets

Task	Steps		See Page
Creating a variable	1	Click 🗐 in the Variable Set Editor toolbar.	300
	2	Type a name for the variable.	
	3	Type a definition for the variable.	
Creating a date/time variable	1	Click 🗐 in the Variable Set Editor toolbar.	301
	2	Type a name for the variable.	
	3	Type a date and time format.	
	4	Select an **Update** option and click **OK**.	
Inserting a variable	1	Select **Insert > Variable**.	301
	2	Select a variable set.	
	3	Select a variable.	
	4	Click **OK**.	
Creating a snippet from existing content	1	Highlight the content you want to convert to a snippet.	308
	2	Select **Home > Create Snippet**.	
	3	Type a name for the snippet.	
	4	Select **Replace Source Content with the New Snippet**.	
	5	Click **Create**.	
Creating a snippet from new content	1	Select **File > New**.	309
	2	For **File Type**, select **Snippet**.	
	3	Select a **Source** template.	
	4	Type a **File Name**.	

Task	Steps	See Page
	5 Click **Add**.	
Inserting a snippet	1 Select **Insert** > **Snippet**. 2 Select a snippet. 3 Click **OK**.	309

Condition tags

Task	Steps	See Page
Creating a condition tag	1 Open a condition tag set. 2 Click 📋 in the Condition Tag Set Editor toolbar. 3 Type a new name for the tag and press **Enter**. 4 Select a color.	314
Applying a condition tag to content	1 Select the content to be tagged. 2 Select **Home** > **Conditions**. 3 Select a condition tag's checkbox. 4 Click **OK**.	315
Applying a condition tag to a topic, file, or folder	1 Select the topic, file, or folder to be tagged. 2 Click 📄 in the Content Explorer toolbar. 3 Select the **Conditional Text** tab. 4 Select a condition tag's checkbox. 5 Click **OK**.	315
Applying a condition tag to a TOC book or page	1 Open the TOC. 2 Select a book or page and click 📄. 3 Select the **Conditional Text** tab. 4 Select a condition tag's checkbox. 5 Click **OK**.	317
Using snippet conditions	1 Apply a condition tag to content in a snippet. 2 Insert the snippet into your topics. 3 Select **View** > **File List**. 4 Right-click the topics that should include the tagged content in the snippet and select **Properties**.	319

Task	Steps	See Page
	5 Select the **Snippet Conditions** tab.	
	6 Set the condition tag to **Include**.	
	7 Click **OK**.	
	8 Select the topics that should exclude the tagged content in the snippet, set the condition tag to **Exclude**, and click **OK**.	

Micro content

Task	Steps	See Page
Creating a micro content file	1 Select **File** > **New**. 2 For **File Type**, select **Micro Content**. 3 Select a **Source** template. 4 Type a **File Name**. 5 Click **Add**.	323
Creating a micro content block	1 Open a micro content file. 2 Click 🔲. 3 Type a search phrase and press **Enter**. 4 Add the response content.	324

Targets

Task	Steps	See Page
Creating a target	1 Select **File** > **New**. 2 For **File Type**, select **Target**. 3 Select a **Source** template. 4 Type a **File Name**. 5 Click **Add**.	333
Building a target	☐ Right-click a target and select **Build**.	342
Viewing a target	☐ Right-click a target and select **View**.	347

Task	Steps	See Page
Creating a publishing destination	1 Select **File** > **New**. 2 For **File Type**, select **Destination**. 3 Select a **Source** template. 4 Type a **File Name**. 5 Click **Add**.	348
Publishing a target	1 Open a target. 2 Click Publish. 3 Select the publishing destination(s). 4 Click **Start Publishing**.	348
Batch generating targets	1 Create and open a batch target. 2 Select the **Schedule** tab. 3 Click **New**. 4 Select a frequency **Setting**. 5 Select a **Start** date and time. 6 If you selected a daily, weekly, or monthly frequency setting, select the recurrence details. 7 If the batch generate should repeat, select **Repeat task every** and specify how often and how long the repeating should occur. 8 If the repeating should expire, select **Expire** and specify an expiration date. 9 If you are ready to enable batch generation, select **Enable**.	369

Context-sensitive help

Task	Steps	See Page
Creating a header file	1 Select **File** > **New**. 2 For **File Type**, select **Header File**. 3 Select a **Source** template. 4 Type a **File Name**. 5 Click **Add**.	375

Task	Steps	See Page
Creating an alias file	1 Select **File** > **New**. 2 For **File Type**, select **Alias File**. 3 Select a **Source** template. 4 Type a **File Name**. 5 Click **Add**.	375
Assigning an identifier to a topic	1 Open an alias file. 2 Select an **Identifier**. 3 Select a **Topic** or **Micro Content** block. 4 Click 🖼.	376
Testing context-sensitive help	1 Build your target. 2 Right-click the target in the Targets folder and select **Test CSH API Calls**. 3 Next to each identifier, click **Test**.	377

Templates

Task	Steps	See Page
Creating a topic template	1 Open the topic that will become the template. 2 Add and format the content you want to include in the template. 3 Select **File** > **Save** > **Save as Template**. 4 Type a **Template Name** and click **OK**.	383
Creating a Contribution template	1 Open the topic that will become the template. 2 Add and format the content you want to include in the template. 3 Select **File** > **Save** > **Save as Contribution Template**. 4 Type a **Template Name**. 5 Select a template location and click **Next**. 6 Select any content files the template uses and click **Next**. 7 Select any condition tags or variable sets the template uses and click **Next**. 8 Click **Finish**.	384

File tags

Task	Steps	See Page
Creating a file tag set	1 Select **File** > **New**. 2 For **File Type**, select **File Tag Set**. 3 Select a **Source** template. 4 Type a **File Name**. 5 Click **Add**.	396
Creating a file tag	1 Open a file tag set. 2 Click 📄. 3 Type a name for the tag.	397
Applying a tag	1 Right-click a file and select **Properties**. 2 Select the **File Tag** tab. 3 Select a **Tag Type**. 4 Select a **File Tag**. 5 Click **OK**.	398

Meta tags Task	Steps	See Page
Creating a meta tag set	1 Select **File** > **New**. 2 For **File Type**, select **Meta Tag Set**. 3 Select a **Source** template. 4 Type a **File Name**. 5 Click **Add**.	400
Creating a meta tag	1 Open a meta tag set. 2 Click ✎. 3 Select Text or List. 4 Type a Name for the meta tag. 5 If you are creating a list meta tag, type a value for the first list item. 6 Click ➕ to add additional list items. 7 If needed, type a default **Value**.	401

Reports

Task	Steps	See Page
Creating an information report	1 Select **File** > **New**. 2 For **File Type**, select **Report File**. 3 Select a **Source** template. 4 Type a **File Name**. 5 Click **Add**.	407
Generating an information report	8 Open a report. 9 On the **General** tab, type a **Title** for the report. 10 Select the **Tables** to include in the report. 11 Click **Generate**.	408

Annotations and topic reviews

Task	Steps	See Page
Adding an annotation	1 Position your cursor where you want to add the annotation. 2 Select **Review** > **Insert Annotation**. 3 If this is your first annotation, type your initials. 4 Type your annotation.	412
Locking content	1 Position your cursor inside the content you want to lock. 2 Select **Review** > **Lock**.	412
Sending a topic for review	1 Select **Review** > **Send for Review**. 2 Type a **Review Package Name**. 3 Click ➕, select the topics you want to send for review, and click **Open**. 4 Click **Next**. 5 Type a subject and message for the email. 6 Select a recipient. 7 Click **Send**.	413

Task	Steps	See Page
Accepting a reviewed topic	1 Select **Review** > **Topic Reviews**. 2 Select **Inbox**. 3 Select a topic. 4 Click ⬚. 5 Click **OK**.	415

Pulse

Task	Steps	See Page
Enabling in a skin	1 Open a skin. 2 Select the **Community** tab. 3 Select **Display topic comments at the end of each topic**. 4 Select **Display Community Search Results**.	421
Enabling in a target	1 Open a target. 2 Select the **Analytics** tab. 3 Select **Enable Pulse/Feedback Server**. 4 Type your Pulse server's **URL**. 5 Click **Login**. 6 Type your **User name**. 7 Type your **Password**. 8 Click **OK**. 9 In the **Communities** field, select your Pulse community.	421
Creating a profile	1 Open an HTML5 or WebHelp target in a browser. 2 Select the **Community** tab or open a topic. 3 Click **Login**. 4 Click **Register**. 5 Type your **First Name** and **Last Name**. 6 Type your **Email Address**. 7 Type and confirm your **Password**. 8 Click **Register**.	422

Task	Steps	See Page
	9 Open your email and complete the registration.	
Editing a profile	1 Open an HTML5 or WebHelp target in a browser.	423
	2 Select the **Community** tab.	
	3 Click **Edit My Profile**.	
Viewing notifications	1 Open an HTML5 or WebHelp target in a browser.	423
	2 Select the **Community** tab.	
	3 Click **Notifications**.	
Viewing reports	1 Open your Pulse Admin site in a browser.	424
	2 Click **Administration**.	
	3 Select **Reports**.	
	4 Click a report.	

Source control

Task	Steps	See Page
Binding a project	1 Select **Project > Project Properties**.	425
	2 Select the **Source Control** tab.	
	3 Click **Bind Project**.	
	4 Select a **Source Control Provider** and provide the required information.	
	5 Select **Keep files checked out** if you want to keep the files checked out.	
	6 Click **OK**.	
Importing a project from source control	1 Select **File > New Project > Import Project**.	427
	2 Select a **Source Control Provider**.	
	3 Select the same server, team project, and other options that were used to bind the project.	
	4 Click **Finish**.	
Adding a file	1 Right-click a file or folder and select **Source Control > Add File**.	428

Task	Steps	See Page
	2 If you want to keep editing the file, select **Keep Checked Out.**	
	3 Click **OK.**	
Getting the latest version	1 Right-click a file or folder and select **Source Control > Get Latest Version.**	429
	2 Click **OK.** If the files are different, click **Merge All** or **Resolve.**	
	3 Click **OK.**	
Checking out a file	1 Right-click a file or folder and select **Source Control > Check Out.**	431
	2 If you want to prevent other users from checking out the file, select **Lock Files.**	
	3 Click **Check Out.**	
Checking in a file	1 Right-click a file or folder and select **Source Control > Check In.**	431
	2 Type a check in **Comment.**	
	3 Select **Keep Checked Out** if you want to check in the current version and keep editing the file.	
	4 Click **Check In.**	
Viewing a list of checked out files	1 Select **Source Control > Pending Check-Ins.**	432
	2 Scroll to the right to view the Status and User columns.	
Viewing differences	1 Right-click a file and select **Source Control > View History.**	432
	2 Select two versions of the file.	
	3 Click **Show Differences.**	
Rolling back	1 Right-click a file and select **Source Control > View History.**	433
	2 Select a version of the file.	
	3 Click **Get Selected Version.**	
Merging changes	1 Check in or get the latest version of a file.	434
	2 Click **Auto Merge All.**	
	3 Click **OK.**	
	4 Click **Resolve.**	
	5 Select a resolution option.	

Task	Steps	See Page
Creating a branch	1 Commit any pending changes on the current branch. 2 Select **Source Control** > **Branch**. 3 Click **Create**. 4 Select a **Source Branch**. 5 Type a **Branch Name**. 6 Click **Create**.	435
Getting a branch	1 Select **Source Control** > **Pull**. 2 Select **Source Control** > **Branch**. 3 Select the **Remotes** tab. 4 Select a branch. 5 Click **Switch**.	435
Switching to a branch	1 Commit any pending changes on the current branch. 2 Select **Source Control** > **Branch**. 3 Select the **Locals** or **Remotes** tab. 4 Select a branch. 5 Click **Switch**.	436
Merging a branch	1 If needed, switch to the branch into which you want to merge. 2 Select **Source Control** > **Merge**. 3 Select the branch you want to merge into the current ("active") branch. 4 Click **OK**.	436
Reverting a branch	1 If needed, switch to the branch you want to revert. 2 Select **Source Control** > **Branch History**. 3 Select the commit you want to revert. 4 Click **Revert**. 5 Click **Accept**.	437
Deleting a branch	1 If you want to delete the current (active) branch, switch to another branch. 2 Select **Source Control** > **Branch**. 3 Select a branch. 4 Click **Delete**.	437

eLearning

Task	Steps	See Page
Creating a multiple choice question	1 Select **eLearning** > **Multiple Choice**. 2 Type the question. 3 Type the first answer. 4 To add more answers, press **Enter** and type the answer. 5 Click the circle before the correct answer.	456
Creating a multiple response question	1 Select **eLearning** > **Multiple Response**. 2 Type the question. 3 Type the first answer. 4 To add more answers, press **Enter** and type the answer. 5 Click the square before each correct answer.	456
Generating an xAPI or SCORM package	1 Open an HTML5 target. 2 Select the **eLearning** tab. 3 Select an eLearning **Standard**. 4 Type a course **Name** and **Description**. 5 Type a unique course **ID**. 6 Select a **Tracking** option. 7 If the Tracking option is set to **Use Course Completion**, type a completion **Percentage**.	463
Creating a test key	1 Open a PDF target. 2 Select the **General** tab. 3 Ensure the Primary TOC is set to the course's TOC. 4 Select the **Advanced** tab. 5 Select **Show correct answers for eLearning questions**. 6 Click **Build**.	464

Answers

Projects

Topics

		Answers
1	C	XHTML is an XML schema.
2	D	You can have as many topics open as you want.
3	A	The topic is linked to an HTML, Word, or FrameMaker document.
4	D	You can link any Flare file between projects, including topics, stylesheets, page layouts, and variables.
5	B	To import a PDF file, you should save it as an HTML and import the HTML file.
6	C	Table styles are table-specific stylesheets, and they have a CSS extension.
7	D	You cannot insert ai files into Flare topics.
8	B	To view a list of topics that contain an image, right-click the image in the Content Explorer and select View Links.

Links

Answers		
1	B	To view a list of topics that link to a topic, right-click the topic in the Content Explorer and select View Links.
2	C	To find and fix broken links, select **Analysis** > **Links** > **Broken Links**.
3	D	Right-click the image in the Content Explorer and select **View Links**.
4	B	You can set a cross reference's link label and formatting using the MadCap|xref style in your stylesheet.
5	A and B	Hyperlinks and popups can open a web page.
6	B	Drop-down links can only show and hide content below the link. Expanding links show and hide content in the same paragraph immediately after the link. Toggler links can show and hide content anywhere in a topic.
7	A	Keyword links display a list of topics that contain a specified index marker.
8	D	Keyword links will not work in the preview. Flare only adds the code for keyword links when you build a target.

Navigation

Answers

1	C	TOC files are stored in the Project Organizer's TOCs folder.
2	A	TOC pages do not have to be inside books.
3	B	Drag the PDF into your TOC.
4	C	You can add second-level index entries using a colon.
5	C	You can select View > Index Window to view your index keywords. Index keywords are stored in your topics, not in an index file.
6	B	To exclude a topic from the search, open the Topic Properties dialog box and deselect the Include topic when full-text search database is generated option.
7	C	Select a **Glossary Term Conversion** option in your online target.
8	B	A browse sequence is an ordered list of links that can be used to find and open topics, like a TOC.

Format and design

1	B	Inline formatting is applied by highlighting content and manually changing its appearance rather than using a style.
2	C	You should consider using a font set for HTML5 if your users are using different operating systems.
3	D	A primary stylesheet is assigned to all topics.
4	D	You can create odd and even pages in a page layout to specify different footers in a print target.
5	C	Specify the page size and margins for a print target.
6	B	A layout that can adjust its width based on the screen width.
7	A	The breadcrumb is the path to the current topic using the TOC.
8	B	To select a skin, open the target and select the skin in the Skin tab.

Single source

1	A	Variables are stored in variable sets in the Project Organizer.
2	B and C	A variable's definition can be set in the VariableSet Editor and in a target on the Variables tab.
3	D	As many as you need.
4	all	Snippets can contain formatted text, tables, lists, and variables.
5	C	Snippets are stored in snippet files in the Content Explorer.
6	all	Condition tags can be applied to topics, folders, TOC books and pages, and index keywords.
7	A	Flare automatically updates the content to use the new condition tag name.
8	D	Condition tags can be used with any target type.

Build and publish

Project management

1	A	Topic templates use the same extension as topics: htm.
2	D	File tags can be applied to any type of file, including topics, images, videos, sounds, page layouts, template pages, skins, and stylesheets.
3	B	Meta tags can be used to include "hidden" information inside the head section of HTML5 topics.
4	C	You can create reports based on file tags and/or unused or used styles, variables, and images.
5	A	Annotations and tracked changes are included when you send a topic for review in Contributor.
6	B	Reviewers can use Contributor to review topics. They do not need to install Flare.
7	C	Flare provides built-in support for Git, Perforce Helix Core, SVN, TFS, and VSS.
8	B	A cloud-based platform for content management, hosting, and task tracking.

Index

CSS to the Point

CSS to the Point provides focused answers to over 200 cascading stylesheet (CSS) questions. Each answer includes a description of the solution, a graphical example, and sample code that has been tested in Chrome, Internet Explorer, Edge, Firefox, Opera, and Safari. If you have been struggling with CSS, this book will help you use CSS like a pro.

You can order *CSS to the Point* at **www.bit.ly/clickstart-books**.

HTML5 to the Point

HTML5 to the Point provides focused answers to over 140 HTML5 questions. Each answer includes a description of the solution and sample code that you can use in your documents. If you want to learn HTML5, this book will help you use it like a pro.

You can order *HTML5 to the Point* at **www.bit.ly/clickstart-books**.

Word 2013 to the Point

Word 2013 to the Point provides answers to over 400 Microsoft Word questions. Each answer includes a description of the solution and step-by-step instructions. The invaluable tips and tricks will help you get started fast, and the comprehensive list of keyboard shortcuts will help you use Word 2013 like a pro.

You can order *Word 2013 to the Point* at **www.bit.ly/clickstart-books**.

Training

ClickStart offers training for Flare, Doc-to-Help, HTML5, CSS, and responsive design. Our training classes extend what you have learned in this book with practice exercises, best practices, and advanced challenges.

We teach online and onsite classes (worldwide), and we offer group discounts for four or more students. For more information, visit our website at **www.clickstart.net** or email us at **info@clickstart.net**.

Consulting

Click**Start** also offers a full range of consulting and contracting services, including:

- ☐ Migrating InDesign, FrameMaker, and Word documents to Flare

- ☐ Migrating Author-it, Doc-to-Help, and RoboHelp projects to Flare

- ☐ Migrating Confluence, Salesforce, and Zendesk content to Flare

- ☐ Developing best practices for creating knowledge bases, online help, user guides, policies and procedures, and training guides

- ☐ Designing stylesheets, page layouts, template pages, skins, and style guides

- ☐ Micro content development, design, and strategy

- ☐ Enhancing the search and integrating external search engines

- ☐ Developing custom features for HTML5 targets

- ☐ Adapting content for use on mobile devices (responsive design)

- ☐ Single sourcing content for multiple audiences

- ☐ Accessibility compliance

- ☐ Publishing content to Confluence, Salesforce, SharePoint, Zendesk, intranets, and the web

- ☐ Developing context-sensitive help, embedded user assistance, microcontent, and microlearning

For more information, visit our website at **www.clickstart.net** or email us at **info@clickstart.net**.